Neumann Systems for the Algebraic AKNS Problem

Recent Titles in This Series

467 **Randolph James Schilling,** Neumann systems for the algebraic AKNS problem, 1992

466 **Shari A. Prevost,** Vertex algebras and integral bases for the enveloping algebras of affine Lie algebras, 1992

465 **Steven Zelditch,** Selberg trace formulae and equidistribution theorems for closed geodesics and Laplace eigenfunctions: finite area surfaces, 1992

464 **John Fay,** Kernel functions, analytic torsion, and moduli spaces, 1992

463 **Bruce Reznick,** Sums of even powers of real linear forms, 1992

462 **Toshiyuki Kobayashi,** Singular unitary representations and discrete series for indefinite Stiefel manifolds $U(p,q;\mathbb{F})/U(p-m,q;\mathbb{F})$, 1992

461 **Andrew Kustin and Bernd Ulrich,** A family of complexes associated to an almost alternating map, with application to residual intersections, 1992

460 **Victor Reiner,** Quotients of coxeter complexes and P-partitions, 1992

459 **Jonathan Arazy and Yaakov Friedman,** Contractive projections in C_p, 1992

458 **Charles A. Akemann and Joel Anderson,** Lyapunov theorems for operator algebras, 1991

457 **Norihiko Minami,** Multiplicative homology operations and transfer, 1991

456 **Michał Misiurewicz and Zbigniew Nitecki,** Combinatorial patterns for maps of the interval, 1991

455 **Mark G. Davidson, Thomas J. Enright and Ronald J. Stanke,** Differential operators and highest weight representations, 1991

454 **Donald A. Dawson and Edwin A. Perkins,** Historical processes, 1991

453 **Alfred S. Cavaretta, Wolfgang Dahmen, and Charles A. Micchelli,** Stationary subdivision, 1991

452 **Brian S. Thomson,** Derivates of interval functions, 1991

451 **Rolf Schön,** Effective algebraic topology, 1991

450 **Ernst Dieterich,** Solution of a non-domestic tame classification problem from integral representation theory of finite groups $(\Lambda = RC_3, v(3) = 4)$, 1991

449 **Michael Slack,** A classification theorem for homotopy commutative H-spaces with finitely generated mod 2 cohomology rings, 1991

448 **Norman Levenberg and Hiroshi Yamaguchi,** The metric induced by the Robin function, 1991

447 **Joseph Zaks,** No nine neighborly tetrahedra exist, 1991

446 **Gary R. Lawlor,** A sufficient criterion for a cone to be area-minimizing, 1991

445 **S. Argyros, M. Lambrou, and W. E. Longstaff,** Atomic Boolean subspace lattices and applications to the theory of bases, 1991

444 **Haruo Tsukada,** String path integral realization of vertex operator algebras, 1991

443 **D. J. Benson and F. R. Cohen,** Mapping class groups of low genus and their cohomology, 1991

442 **Rodolfo H. Torres,** Boundedness results for operators with singular kernels on distribution spaces, 1991

441 **Gary M. Seitz,** Maximal subgroups of exceptional algebraic groups, 1991

440 **Bjorn Jawerth and Mario Milman,** Extrapolation theory with applications, 1991

439 **Brian Parshall and Jian-pan Wang,** Quantum linear groups, 1991

438 **Angelo Felice Lopez,** Noether-Lefschetz theory and the Picard group of projective surfaces, 1991

437 **Dennis A. Hejhal,** Regular b-groups, degenerating Riemann surfaces, and spectral theory, 1990

436 **J. E. Marsden, R. Montgomery, and T. Ratiu,** Reduction, symmetry, and phase mechanics, 1990

(Continued in the back of this publication)

Number 467

Neumann Systems for the Algebraic AKNS Problem

Randolph J. Schilling

May 1992 • Volume 97 • Number 467 (first of 3 numbers) • ISSN 0065-9266

American Mathematical Society
Providence, Rhode Island

1991 *Mathematics Subject Classification.*
Primary 58F07; Secondary 35Q20, 58F05, 58F19, 17B65, 14H40.

Library of Congress Cataloging-in-Publication Data

Schilling, Randolph J. (Randolph James), 1951–
Neumann systems for the algebraic AKNS problem/Randolph J. Schilling.
p. cm. – (Memoirs of the American Mathematical Society, ISSN 0065-9266; no. 467)
Includes bibliographical references.
ISBN 0-8218-2537-2
1. Hamiltonian systems. 2. Evolution equations. 3. Jacobians. I. Title. II. Title: AKNS problem. III. Series.
QA3.A57 no. 467
[QA614.83]
510 s–dc20 92-6951
[514′.74] CIP

Memoirs of the American Mathematical Society

This journal is devoted entirely to research in pure and applied mathematics.

Subscription information. The 1992 subscription begins with Number 459 and consists of six mailings, each containing one or more numbers. Subscription prices for 1992 are $292 list, $234 institutional member. A late charge of 10% of the subscription price will be imposed on orders received from nonmembers after January 1 of the subscription year. Subscribers outside the United States and India must pay a postage surcharge of $25; subscribers in India must pay a postage surcharge of $43. Expedited delivery to destinations in North America $30; elsewhere $82. Each number may be ordered separately; *please specify number* when ordering an individual number. For prices and titles of recently released numbers, see the New Publications sections of the *Notices of the American Mathematical Society*.

Back number information. For back issues see the *AMS Catalogue of Publications*.

Subscriptions and orders should be addressed to the American Mathematical Society, P. O. Box 1571, Annex Station, Providence, RI 02901-1571. *All orders must be accompanied by payment.* Other correspondence should be addressed to Box 6248, Providence, RI 02940-6248.

Memoirs of the American Mathematical Society is published bimonthly (each volume consisting usually of more than one number) by the American Mathematical Society at 201 Charles Street, Providence, RI 02904-2213. Second-class postage paid at Providence, Rhode Island. Postmaster: Send address changes to Memoirs, American Mathematical Society, P. O. Box 6248, Providence, RI 02940-6248.

10 9 8 7 6 5 4 3 2 1 97 96 95 94 93 92

TABLE OF CONTENTS

Abstract. The Neumann system consists of harmonic oscillators constrained to move on the unit sphere in configuration space. It is an algebraically completely integrable Hamiltonian system. It is known that any finite gap potential of Hill's equation may be expressed in terms of a solution to the Neumann problem. This paper is concerned with an algebraically completely integrable Hamiltonian system whose solutions may be used to describe the finite gap solutions of the AKNS spectral problem, a first order 2×2 matrix linear system. Trace formulas, constraints, Lax Pairs and constants of motion are obtained using Krichever's algebraic inverse spectral transform. Computations are carried out explicitely over the class of spectral problems with $\ell \times \ell$ matrix coeficients.

Key Words: Neumann system, complete integrability, reduction, momentum mapping, loop algebra, Riemann surface, Baker function.

Acknowledgements: I want to thank Henry McKean Jr. and Hermann Flaschka for productive discussions throughout the development of this paper.

INTRODUCTION

Background. This paper is concerned with a hierarchy of nonlinear evolution equations which includes the nonlinear Schrödinger equation,

$$q_t = \frac{i}{2} q_{xx} - i|q^2|q, \tag{0.1}$$

and the modified Korteweg-de Vries equation,

$$q_t = -\frac{1}{4} q_{xxx} - \frac{3}{2} q^2 q_x. \tag{0.2}$$

These equations are ζ preserving deformations of the following linear spectral problem:

$$\begin{cases} \dot{\psi}_1 = -i\zeta\psi_1 + q(x)\psi_2 \\ \dot{\psi}_2 = r(x)\psi_1 + i\zeta\psi_1 \end{cases} \tag{0.3a}$$

or

$$\dot{\psi} = (\epsilon\zeta + \mathbf{q}(x))\psi \quad \text{where} \quad \epsilon = \operatorname{diag}(-i, i), \tag{0.3b}$$

$\psi^T = (\psi_1, \psi_2)$ and $\mathbf{q}$ is a diagonal free 2×2 matrix. There are several methods for solving these equations. The infinite line problem, $-\infty < x < \infty$ with $q(x) \to 0$ as $|x| \to \infty$, was solved in the paper by Ablowitz, Kaup, Newell and Segur (AKNS) [1] using inverse scattering. The periodic problem may be solved using Floquet theory and the periodic inverse spectral transform [13].

Finite Dimensional Invariant Surfaces. Krichever [12] (See also [15].) used algebraic techniques to expess certain solutions in terms of θ functions. He was concerned with finite dimensional invariant spaces of solutions. His approach is based on the following fact. If $q(x,t)$ satisfies (0.1), (0.2) or any equation of the

AKNS hierarchy, and if there exists a differential operator P such that $q(x,0)$ satisfies

$$(0.4) \qquad [L,P] \doteq LP - PL = 0,$$

where

$$(0.5) \qquad L = \epsilon^{-1}(\partial - \mathbf{q}), \quad \partial = \cdot = \frac{\partial}{\partial x},$$

then $q(x,t)$, as a function of x, also satisfies (0.4). The equation (0.4), which says that the differential operators L and P commute, disguises a nonlinear system of ordinary differential equations. It is known that commuting differential operators satisfy an algebraic equation of the form $f(x,y)=0$. The algebraic curve $f(x,y)=0$ is called the *spectral curve*. The dimension of the system is directly related to the order of P and to the genus of the spectral curve. The space of solutions to (0.4), which is finite dimensional, is invariant under the equations of the AKNS hierarchy.

AKNS-Neumann System. This paper is concerned with finite dimensional invariant spaces of solutions but the approach does not use commuting differential operators. The approach is based on the following system of equations, which is called the AKNS-Neumann system,

$$(0.6a) \qquad \dot{\mathbf{m}} = X(\mathbf{m}) : \begin{cases} \dot{x}_1 = & \epsilon_1 \mathbf{a} x_1 + (x_1 \cdot u_2) x_2 \\ \dot{x}_2 = & \epsilon_2 \mathbf{a} x_2 + (x_2 \cdot u_1) x_1 \\ \dot{u}_1 = & -\epsilon_1 \mathbf{a} u_1 - (x_2 \cdot u_1) u_2 \\ \dot{u}_2 = & -\epsilon_2 \mathbf{a} u_2 - (x_1 \cdot u_2) u_1 \end{cases}$$

where $x_1, x_2, u_1, u_2 \in \mathbb{C}^{\ell'}$, $\mathbf{m} \doteq (x_1, x_2, u_1, u_2)$, $\epsilon_1 = i$, $\epsilon_2 = -i$ and $\mathbf{a}$ is a constant diagonal matrix with distinct entries a_r. Here ℓ' is an integer ≥ 1. The system (0.6a), more specifically, a reduced version of (0.6a), is used to describe the polymerization of certain plastics [5]. Here, τ is used instead of x as the independent variable. The equations have the following equivalent matrix formulation:

$$(0.6b) \qquad \begin{cases} \dot{x}^r = (a_r \epsilon + \mathbf{q}(\tau)) x^r \\ \dot{u}^r = -(a_r \epsilon + \mathbf{q}(\tau))^T u^r \end{cases} \quad \text{where} \quad x^r = \begin{pmatrix} x_1^r \\ x_2^r \end{pmatrix}, \quad u^r = \begin{pmatrix} u_1^r \\ u_2^r \end{pmatrix},$$

$\mathbf{q}$ is given by the following formula:

$$(0.7) \qquad \mathbf{q} = \sum_{r=1}^{\ell'} x^r \otimes u^r - z_1 = \mathbf{x}^T \mathbf{u} - z_1,$$

$z_1 = \text{diag}(x_1 \cdot u_1,\, x_2 \cdot u_2)$, $\mathbf{x}$ is the $\ell' \times 2$ matrix (x_1, x_2), $\mathbf{u} = (u_1, u_2)$ and $x^r \otimes u^r$ is the 2×2 matrix whose entry (i,j) is $x_i^r u_j^s$. The x^r and u^r satisfy these equations:

$$Lx^r = a_r x^r \quad \text{and} \quad L^\dagger u^r = a_r u^r$$

where L is as in (0.5) with $\mathbf{q}$ defined by (0.7). The centralizer of L is nontrivial and the genus of the underlying spectral curve is $g = \ell' - 1$. In Chapter II, using Baker functions and the residue theorem for Riemann surfaces, it is shown that a generic class of commutative rings may be constructed in this way using (0.6a). This is done for first order $\ell \times \ell$ matrix differential operators L generalizing the 2×2 AKNS problem.

One advantage of this approach to finite dimensional invariant sets is its simplicity. The complexity and the appearance of the equations (0.4) is genus dependent; for large genus, (0.4) is complicated. The phase space is a differential algebra and the underlying Hamiltonian mechanics is nonclassical. The equations have not been written explicitely except for some low genus problems. The genus of the curve underlying the AKNS-Neumann system is the parameter g so the appearance of the equations does not depend in any way on the genus. The Hamiltonian mechanics, which is described in Chapter IV, is classical.

Isospectral Matrices. A Hamiltonian vector field on a manifold M of dimension $2n$ is called *completely integrable* if there is a set of n independent Hamiltonian vector fields on M that commute with each other and the original Hamiltonian vector field. The Hamiltonians are constants of motion and the Poisson bracket of any two of the Hamiltonians is zero. The Hamiltonians are said to be *in involution.* One way of finding constants of motion is to find an operator that is *isospectral* with respect to the system of equations. An matrix Z is said to be *isospectral* with respect to a vector field X if there exists an matrix A such that $X(Z) = [A, Z]$. It follows then that the characteristic polynomial of Z and the functions $\text{Tr}(Z^j)$ are constants of motion. In chapter II the following matrices are derived and shown to be isospectral along a particular solution to the AKNS-Neumann problem (0.6a):

$$Z = \epsilon + \sum_{r=1}^{\ell'} \frac{x^r \otimes u^r}{\zeta - a_r} = \epsilon + \mathbf{x}^T (\zeta - \mathbf{a})^{-1} \mathbf{u} \tag{0.8}$$

and

$$\mathbb{L} = \mathbf{a} + \sum_{\alpha=1}^{\ell} \frac{x_\alpha \otimes u_\alpha}{z - \epsilon_\alpha} = \mathbf{a} + \mathbf{x}(z-\epsilon)^{-1}\mathbf{u}^T \tag{0.9}$$

They are shown to be globally isospectral in Chapter III. The characteristic polynomial of Z can be expanded in a fairly straight forward manner, its coefficients can be differentiated along (0.6a) and shown to be independent of x. These direct but tedious computations are circumvented in Chapter III.

Each equation of the AKNS hierarchy is the integrability condition of a pair of linear equations of the this form:

$$\dot{\psi} = (\epsilon\zeta + \mathbf{q}(x,t))\psi \quad \text{and} \quad \frac{\partial \psi}{\partial t}(x,t) = B(\mathbf{q}(x,t),\zeta)\psi(x,t). \tag{0.10}$$

The second equation, upon restriction to the AKNS-Neumann variables $\mathbf{m} = (\mathbf{x}, \mathbf{u})$, leads to the following system on $\mathbb{C}^{2\ell\ell'}$:

$$\dot{x}^r = B(\mathbf{q}(\tau,t), a_r)x^r \quad \text{and} \quad \dot{u}^r = -B(\mathbf{q}(\tau,t), a_r)^T u^r. \tag{0.11}$$

The entries of B are known to be differential polynomials in the entries of $\mathbf{q}$ and as such, the entries of B are functions of $\mathbf{m}$. Therefore the equations (0.11) define a vector field. These vector fields make up the AKNS-Neumann hierarchy. Hamiltonian are obtained from traces of powers of $\mathbb{L}$ in Chapter IV. The geometry of momentum mappings and the Adler-Kirrillov-Kostant-Symes theorem are used in Chapter IV to prove that the Hamiltonians are in involution. It would be extremely difficult to prove by direct computation that there are g independent constants of motion among those coming from the characteristic polynomial. This computation is circumvented in Chapter III using a correspondence between points $\mathbf{m}$ and divisors over Riemann surfaces called the divisor map. The vector field X is, at any rate, a completely integrable Hamiltonian system.

A system of equations is called *algebraically completely integrable* [4c] if its solutions are expressible in terms of Abelian function. The solution to (0.6a) is given in terms of θ functions in Chapter II. It follows from this construction that the AKNS-Neumann system is algebraically completely integrable.

The differential geometry for systems of Neumann type was developed in [3]. See also [16] and [17]. Chapter I contains background material on such matters.

It includes Lie algebraic and group theoretical preliminaries such as the Adler-Kirrillov-Kostant-Symes theorem. It contains an involution theorem which is used in Chapter IV to prove that the constants of motion of the AKNS-Neumann problem are in involution. Adams-Harnad-Previato applied this background and the integrability theorems in [17] to certain rank $\ell - 1$ generalizations of the Neumann systems. (A Neumann system is said to have rank s if the multiplicity of each eigenvalue of $\mathbf{a}$ is s.) Their results are not applicable to the AKNS-Neumann systems which are all of rank 1.

The generalities of the introduction are illustrated by the the Neuman system itself in the appendix.

CHAPTER I

THE GEOMETRY OF NEUMANN SYSTEMS

A *Hamiltonian structure* on a finite dimensional C^∞-manifold M is an antisymmetric bilinear form:

$$\{,\} : C^\infty(M) \times C^\infty(M) \to C^\infty(M)$$

such that for all f,g,h $\in C^\infty(M)$ one has

$$\begin{cases} \{f, gh\} = \{f, g\}h + g\{f, h\} \\ \{f, \{g, h\}\} = \{\{f, g\}, h\} + \{g, \{f, h\}\}. \end{cases}$$

The second of these formulas is used to show that $C^\infty(M)$ is a Lie algebra with Lie bracket $\{,\}$. The first formula is used to show that the mapping:

$$h \in C^\infty(M) \to D_h : f \in C^\infty(M) \to D_h(f) = \{f, h\} \in C^\infty(M)$$

is a homomorphism of $C^\infty(M)$ into the set of vector fields on M; D_h is called the *Hamiltonian vector field* of h.

The gradient of a function $f \in C^\infty(M)$ is the section of the cotangent bundle $T^*(M)$ given for $p \in M$ and $\xi \in T_p(M)$ by these formulas:

$$df(p) = T_p f : T_p(M) \to \mathbb{C}, \quad < df(p), \xi >= \frac{d}{dt} f(\gamma(t)) \mid_{t=0}$$

where γ is a curve leaving p with velocity ξ. We shall often use the notation ∇f in place of df.

The Kirrillov/Poisson structure $(\mathfrak{g}^*, \{,\}_{\mathfrak{g}^*})$ on the dual $\mathfrak{g}^*$ of a Lie algebra $\mathfrak{g}$ is given for $f, g \in C^\infty(\mathfrak{g}^*)$ and $\alpha \in \mathfrak{g}^*$ by this formula:

$$\{f, g\}_{\mathfrak{g}^*}(\alpha) = \alpha([\nabla f(\alpha), \nabla g(\alpha)]).$$

Received by the editors January 10, 1989.
Supported by NSF MCS 8211308.

We have used these standard identifications: $T^*_\alpha(\mathfrak{g}^*) \cong \mathfrak{g}^{**} \cong \mathfrak{g}$. The Hamiltonian vector field of $h \in C^\infty(\mathfrak{g}^*)$ is given at $\alpha \in \mathfrak{g}^*$ by this formula:

$$D_h(\alpha) = -\operatorname{ad}^*_{\nabla h(\alpha)}(\alpha).$$

Suppose our Lie algebra admits a symmetric nondegenerate bilinear form $K : \mathfrak{g} \times \mathfrak{g} \to \mathbb{C}$ satisfying the associativity law: $K([\xi, \eta], \rho) = K(\xi, [\eta, \rho])$ for all $\xi, \eta, \rho \in \mathfrak{g}$. We shall identify the dual of $\mathfrak{g}$ with $\mathfrak{g}$ itself using K in the usual way. Suppose there is given a direct sum decomposition of $\mathfrak{g}$ by subalgebras $\mathfrak{K}$ and $\mathfrak{N}$: $\mathfrak{g} = \mathfrak{N} \oplus \mathfrak{K}$. Let $\pi_{\mathfrak{N}}$ denote the projection onto $\mathfrak{N}$ along $\mathfrak{K}$. Let $\mathfrak{K}^0$ denote the annihilator of $\mathfrak{K}$ with respect to K. Let $\pi_{\mathfrak{K}^0}$ denote the projection onto $\mathfrak{K}^0$ along $\mathfrak{N}^0$. Then $\mathfrak{K}^0$ is a subspace of $\mathfrak{g}$ that is not, in general, a Lie subalgebra (unless $\mathfrak{K}$ is an ideal). If $\epsilon \in \mathfrak{g}$ then there is a natural linear isomorphism: $\psi_\epsilon : \epsilon + \mathfrak{K}^0 \to \mathfrak{N}^*$. The pull back $\{,\}_\epsilon$ of the Hamiltonian structure $\{,\}_{\mathfrak{N}^*}$ under ψ_ϵ is given for $f, g \in C^\infty(\epsilon + \mathfrak{K}^0)$ and $\xi \in \mathfrak{K}^0$ by this formula:

(1.1a) $$\{f, g\}_\epsilon(\epsilon + \xi) = K(\xi, [\pi_{\mathfrak{N}} \circ \nabla f(\epsilon + \xi), \pi_{\mathfrak{N}} \circ \nabla g(\epsilon + \xi)]).$$

The Hamiltonian vector field corresponding to an h in $C^\infty(\epsilon + \mathfrak{K}^0)$ is given by this formula:

(1.1b) $$D_h(\epsilon + \xi) = \pi_{\mathfrak{K}^0}[\pi_{\mathfrak{N}} \circ \nabla h(\epsilon + \xi), \xi].$$

A function $f \in C^\infty(\mathfrak{g})$ is said to be *invariant* if it satisfies $[\nabla f(\xi), \xi] = 0$ for all $\xi \in \mathfrak{g}$.

(1.2) THEOREM: (Adler/Kostant/Symes): Let us choose an *infinitesimal character* ϵ of $\mathfrak{N}$; i.e., ϵ belongs to the subspace $\mathfrak{N}^0 \cap [\mathfrak{K}, \mathfrak{K}]^0$.

(1.2a) The Hamiltonian vector field of the restriction to $\epsilon + \mathfrak{K}^0$ of an invariant $h \in C^\infty(\mathfrak{g})$ is given by this formula:

$$D_h(\epsilon + \xi) = -[\pi_{\mathfrak{K}} \circ \nabla h(\epsilon + \xi), \epsilon + \xi] = [\pi_{\mathfrak{N}} \nabla h(\epsilon + \xi), \epsilon + \xi], \quad \xi \in \mathfrak{K}^0.$$

(1.2b) The set of all restictions to $\epsilon + \mathfrak{K}^0$ of invariant functions on $\mathfrak{g}$ is involutive with respect to $\{,\}_\epsilon$.

Proof: If f and g are restrictions to $\epsilon + \mathfrak{K}^0$ of invariant functions and $\xi \in \mathfrak{K}^0$ then one has

$$\begin{aligned}\{f,g\}_\epsilon(\epsilon+\xi) &= K(\xi, [\pi_{\mathfrak{N}} \circ \nabla f(\epsilon+\xi), \pi_{\mathfrak{N}} \circ \nabla g(\epsilon+\xi)]) \\ &= K(\epsilon+\xi, [\pi_{\mathfrak{N}} \circ \nabla f(\epsilon+\xi), \pi_{\mathfrak{N}} \circ \nabla g(\epsilon+\xi)]) \\ &= K(\epsilon+\xi, [\pi_{\mathfrak{K}} \circ \nabla f(\epsilon+\xi), \pi_{\mathfrak{K}} \circ \nabla g(\epsilon+\xi)]) \\ &= 0.\end{aligned}$$

The first equality is just the definition (1.1a); the second equality follows from $\epsilon \in \mathfrak{N}^0$; the third equality follows from $1 = \pi_{\mathfrak{N}} + \pi_{\gamma K}$, the associativity of K and the invariance of f and g; the last equality follows from $\epsilon \in [\mathfrak{K}, \mathfrak{K}]^0$, $\xi \in \mathfrak{K}^0$ and the fact that $\mathfrak{K}$ is a subalgebra.

If h is the restriction to $\epsilon + \mathfrak{K}^0$ of an invariant function then for $\xi \in \mathfrak{K}^0$ one has

$$\begin{aligned}D_h(\epsilon+\xi) &= \pi_{\mathfrak{K}^0}[\pi_{\mathfrak{N}} \circ \nabla h(\epsilon+\xi), \xi] \\ &= \pi_{\mathfrak{K}^0}[\pi_{\mathfrak{N}} \circ \nabla h(\epsilon+\xi), \epsilon+\xi] \\ &= -\pi_{\mathfrak{K}^0}[\pi_{\mathfrak{K}} \circ \nabla h(\epsilon+\xi), \epsilon+\xi] \\ &= -[\pi_{\mathfrak{K}} \circ \nabla h(\epsilon+\xi), \epsilon+\xi] \\ &= [\pi_{\mathfrak{N}} \circ \nabla h(\epsilon+\xi), \epsilon+\xi].\end{aligned}$$

All the equalities except the fourth one follow from reasons cited above. Using the associativity of K, one may show that the commutator in the fourth equation already belongs to $\mathfrak{K}^0$. □

There is an often used alternative version of theorem (1.2) in which, roughly speaking, everything is translated back to $\mathfrak{K}^0$. We would like to state this version just to avoid confusion.

(1.3) THEOREM: Let us suppose that ϵ belongs to $\mathfrak{N}^0 \cap [\mathfrak{K}, \mathfrak{K}]^0$. Let

$$\Upsilon_\epsilon = \{f \circ S_\epsilon \circ \iota : f \text{ is an invariant function in } C^\infty(\mathfrak{g})\}$$

where S_ϵ denotes translation in $\mathfrak{g}$ by ϵ and $\iota : \mathfrak{K}^0 \to \mathfrak{g}$ is the inclusion map.

(1.3a) The Hamiltonian vector field of $h \circ S_\epsilon \circ \iota$ i s given for ξ in $\mathfrak{K}^0$ by this formula:

$$D_{h\circ S_\epsilon \circ \iota}(\xi) = -[\pi_{\mathfrak{K}} \circ \nabla h(\epsilon + \xi), \epsilon + \xi] = [\pi_{\mathfrak{N}} \circ \nabla h(\epsilon + \xi), \epsilon + \xi].$$

(1.3b) The set Υ_ϵ is involutive with respect to $\{,\}_0$.

Proof. This version follows from the formula $\nabla h \circ S \circ \iota(\xi) = \nabla h(\epsilon + \xi)$ and the arguements used in the previous proof. □

Let G be a Lie group with Lie algebra $\mathfrak{g}$. Let (M, ω) be a symplectic manifold. Let $\{,\}$ be the Poisson bracket associated to ω and let X_H denote the Hamiltonian vector field of an H in $C^\infty(M)$. Suppose there is given a Lie group action Φ : $G \times M \to M$. The G-action is said to be Hamiltonian if the following three conditions are satisfied:

(1.4a) For each $g \in G$ the mapping $\Phi_g : M \to M$, the projection of $\Phi \mid_{\{g\}\times M}$ onto the second argument, is a canonical transformation; that is, $\Phi_g^* \omega = \omega$.

(1.4b) Let $\Phi_{,p}, p \in M$ denote the projection of $\Phi \mid_{G\times\{p\}}$ onto the first factor. For each $\xi \in \mathfrak{g}$ the induced vector field $\xi_M(p) = T_e \Phi_{,p} \cdot \xi$ is Hamiltonian with respect to ω; that is, there exists an $F_\xi \in C^\infty(M)$ such that $\omega(\xi_M, \cdot) = dF_\xi$.

(1.4c) The mapping $\xi \to F_\xi$ is an homomorphism of Lie algebras that is *equivariant* in the sense that $F_{\mathrm{Ad}_g(\xi)} = F_\xi \circ \Phi_g$ for all $\xi \in \mathfrak{g}$ and all $g \in G$.

A *momentum mapping* associated to a Hamiltonian G-action is a mapping

$$J : M \to \mathfrak{g}^* \quad \text{such that} \quad J(p) \cdot \xi = F_\xi(p) \quad \forall p \in M \quad \text{and} \quad \xi \in \mathfrak{g}^*.$$

Any momentum mapping is equivariant with respect to the coadjoint action of G on $\mathfrak{g}^*$ and the G-action on M; that is, $J \circ \Phi_g = Ad^*_{g^{-1}} \circ J$ for all $g \in G$. This property follows from (1.4c). A *collective motion* is an integral curve of a Hamiltonian vector field on M with Hamiltonian of the form $h \circ J$ where $h \in C^\infty(\mathfrak{g}^*)$.

(1.5) THEOREM: (Guillemin/Sternberg [11]) Suppose there be given a Hamiltonian G-action on M and let $J : M \to \mathfrak{g}^*$ is an Ad^*-equivariant momentum mapping. Then J is a canonical mapping of Poisson manifolds; that is,

$$\text{(1.5a)} \qquad J_* X_{h\circ J}(J(p)) = D_h(J(p)) \quad \text{for every} \quad p \in M \quad \text{and} \quad h \in C^\infty(\mathfrak{g}^*).$$

Equivalently, the induced mapping $J^* : C^\infty(\mathfrak{g}^*) \to C\infty(M)$ is a homomorphism of Lie algebras,

$$\{f,h\}_{\mathfrak{g}^*} \circ J = \{f \circ J, h \circ J\} \quad \text{for every} \quad f,h \in C^\infty(\mathfrak{g}^*). \tag{1.5b}$$

Proof: If $\xi \in \mathfrak{g}$ then let L_ξ denote the corresponding element of $\mathfrak{g}^{**} \subset C^\infty(\mathfrak{g}^*)$. Then the infinetesimal action of G on M and on $\mathfrak{g}^*$ is given by

$$\xi_M = X_{L_\xi \circ J} \quad \text{and} \quad \xi_{\mathfrak{g}^*} = D_{L_\xi}.$$

One has

$$\begin{aligned}(J_*\xi_M)(J(p)) &= T_0(J \circ \exp(t\xi) \cdot p)\\ &= T_0(\exp(t\xi) \cdot J(p))\\ &= \xi_{\mathfrak{g}^*}(J(p));\end{aligned}$$

hence, we have proven this linear version of (1.5a):

$$J_* X_{L_\xi \circ J} = D_{L_\xi}. \tag{1.5c}$$

Given $h \in C^\infty(\mathfrak{g}^*)$ and $p \in M$ we let $\xi = \nabla h(J(p))$. Then from the formula

$$\xi_M(p) = X_{h \circ J}(p)$$

it is clear that $\exp(t\xi) \cdot p$ is a curve leaving p with velocity $X_{h\circ J}(p)$. This proves that one may replace L_ξ by h in (1.5c) which proves (1.5a).

Now to prove (1.5b) we observe, using (1.5a), that

$$\{f,h\}_{\mathfrak{g}^*}(J(p)) = (D_h(f))(J(p)) = (J_* X_{h\circ J}(f))(J(p))$$

$$= (X_{h\circ J}(f \circ J))(p) = \{f \circ J, h \circ J\}(p)$$

for all $f, h \in C^\infty(\mathfrak{g}^*)$. □

This theorem has some immediate consequences. The collective motion p(t), p(0)=p, of $h \circ J$ is given by

$$p(t) = g(t) \cdot p = \Phi_p(g(t))$$

where g(t) is a curve in G satisfying this system of equations:

$$\dot{g}(t) = T_e R_{g(t)} \cdot \nabla h(J(g(t) \cdot p)) \quad \text{and} \quad g(0) = e$$

where R denotes the right action of G on itself. Thus p(t) lies on the G-orbit of p and if H is any G-invariant Hamiltonian on M then $h \circ J$ is constant along any integral curve of X_H; hence, $\{H, h \circ J\} = 0$. The curve J(p(t)) is an integral curve of D_h.

If $h \in C^\infty(\mathfrak{g}^*)$ is an invariant function then the Hamiltonian vector field of h is trivial and h is a central element of the Lie algebra $(C^\infty(\mathfrak{g}^*), \{,\}_{\mathfrak{g}^*})$. The collective motion of an invariant h is given by

$$p(t) = g(t) \cdot p \quad \text{where} \quad g(t) = \exp(t(\nabla h(J(p)).$$

Let there be given a Hamiltonian G-action on M with momentum mapping J. Let ν be a regular element of $\mathfrak{g}^*$. Then $J^{-1}(\nu)$ is a submanifold of M. The Ad^*-isotropy subgroup of ν, denoted G_ν, acts on $J^{-1}(\nu)$. Let $M_\nu = J^{-1}(\nu)/G_\nu$ denote the set of orbits of this action. If the action is free and proper then M_ν is a manifold and the canonical mapping $\pi_\nu : J^{-1}(\nu) \to M_\nu$ is a submersion ([2], Proposition (4.1.23)). Moreover, there exists a unique symplectic form ω_ν on M_ν such that $\pi_\nu^* \omega_\nu = \iota_\nu^* \omega$ where $\iota_\nu : J^{-1}(\nu) \to M$ is the inclusion mapping. The symplectic manifold (M_ν, ω_ν) is called the (Marsden/Weinstein) *reduced phase space.*

AKNS/Neumann Systems. Let $\mathcal{M}$ denote the cotangent bundle to the space of all $\ell' \times \ell$ matrices. Suppose $\mathbf{x}$ is an $\ell' \times \ell$ matrix with rows $(x^r)^T$ and columns x_α:

$$\mathbf{x} = (x^1, \ldots, x^{\ell'})^T = (x_1, \ldots, x_\ell).$$

An element of $\mathcal{M}$ has the form $\mathbf{m} = (\mathbf{x}, \mathbf{u})$ where $\mathbf{u}$ has rows $(u^r)^T$ and columns u_α. Ths symplectic form is this:

$$\omega_{\mathbf{m}} = \operatorname{tr}(d\mathbf{x} \wedge d\mathbf{u}^T) = \sum_{r=1}^{\ell'} \sum_{\alpha=1}^{\ell} dx_\alpha^r \wedge du_\alpha^r.$$

Let $\mathcal{M}^1$ denote the subset of $\mathcal{M}$ consisting of all $\mathbf{m}$ in which each x^r has rank 1 and each u^r has rank 1. Let $\mathcal{M}_\ell$ denote the subspace of $\mathcal{M}$ consisting of all points $\mathbf{m}$ with $\mathbf{x}$ and $\mathbf{u}$ of rank ℓ. These are dense open submanifolds of $\mathcal{M}$.

Let $\mathfrak{g} = \mathrm{gl}(\ell, \mathbb{C})$, the Lie algebra of all $\ell \times \ell$ matrices over $\mathbb{C}$. We shall identify the dual of $\mathfrak{g}$ with $\mathfrak{g}$ itself using the usual trace form: $(\xi, \eta) \in \mathfrak{g} \times \mathfrak{g} \to \mathrm{Tr}(\xi\eta)$. The *loop algebra* $L(\mathfrak{g})$ in a parameter ζ is the infinite dimensional Lie algebra defined by the following formulas:

$$L(\mathfrak{g}) = \{\, \xi(\zeta) = \zeta^N (\sum_{j=0}^{\infty} \xi_j \zeta^{-j}) : \xi \in \mathfrak{g} \quad \text{and} \quad N \in \mathbb{Z}\}$$

$$[\xi(\zeta), \xi'(\zeta)] = \zeta^{N+N'} \sum_{j=0}^{\infty} \left(\sum_{l+k=j} [\xi_j, \xi'_k] \right) \zeta^{-j}.$$

Let K denote the bilinear form on $L(\mathfrak{g})$ given for $\xi(\zeta), \eta(\zeta) \in L(\mathfrak{g})$ by this formula:

$$K(\xi(\zeta), \eta(\zeta)) = \mathrm{tr}(\pi_{-1}(\xi(\zeta)\eta(\zeta))) = -\operatorname{Res}_\infty \mathrm{tr}(\xi(\zeta)\eta(\zeta))d\zeta.$$

where π_j is the projection taking $L(\mathfrak{g})$ onto the subspace $\mathfrak{g}\zeta^j$ along its natural complement. Then K is symmetric, nondegenerate and associative. Let us consider the direct sum decomposition of $L(\mathfrak{g})$ given by this formula: $L(\mathfrak{g}) = \mathfrak{P} \oplus \mathfrak{Q}$ where $\mathfrak{P}$ is the subalgebra of polynomial loops and $\mathfrak{Q}$ is the complimentary subalgebra of loops in ζ^{-1}. Then, in terms of K, one has

$$\mathfrak{P}^* \cong \mathfrak{Q}^0 = \mathfrak{Q} \quad \text{and} \quad \mathfrak{Q}^* \cong \mathfrak{P}^0 = \mathfrak{P}.$$

The identification of $\mathfrak{P}^*$ with $\mathfrak{Q}^0$ is given by these formulas:

$$\alpha \in \mathfrak{P}^* \to \xi_\alpha(\zeta) \in \mathfrak{Q}^0 : \begin{cases} \alpha(\eta(\zeta)) = K(\xi_\alpha(\zeta), \eta(\zeta)) \\ \text{and} \\ -\mathrm{ad}^*_{\rho(\zeta)}(\alpha) \to -\pi_{\mathfrak{Q}^0}[\xi_\alpha(\zeta), \rho(\zeta)]. \end{cases}$$

The space of infinitesimal characters of $\mathfrak{P}, \mathfrak{P}^0 \cap [\mathfrak{Q}, \mathfrak{Q}]^0$, is $\mathfrak{g}$ itself. Let ϵ be a diagonal element of $\mathfrak{g}$ with distinct entries. We shall be concerned with the Poisson structure of $\mathfrak{P}^*$ pulled back to $\epsilon + \mathfrak{Q}^0$.

Let $\mathfrak{h} = \mathrm{gl}(\ell', \mathbb{C})$ and let $L(\mathfrak{h})$ be its loop algebra with parameter z. The previous dicussion concerning $L(\mathfrak{g})$, trace forms, decomposition $L(\mathfrak{g}) = \mathfrak{P}' \oplus \mathfrak{Q}'$, applies equally well to $L(\mathfrak{h})$. the components of the z-analogue of Let **a** be a diagonal matrix with distinct entries; an infinitesimal character of the decomposition of $L(\mathfrak{h})$.

(1.6) Hamiltonian $\mathbf{G} \doteq \mathbf{GL}(\ell, \mathbb{C})$ Actions. Let $\mathbf{G}^{\ell'}$ denote the ℓ' fold direct product of $\mathbf{G}$ with itself. Then $\mathbf{G}^{\ell'}$ is a Lie group whose Lie algebra may be identified with the ℓ' fold direct sum of $\mathfrak{g}$ with itself. Let $\mathbf{G}_\epsilon$ be the isotropy subgroup of ϵ. It is the group of invertible diagonal matrices and its Lie algebra $\mathfrak{g}_\epsilon$ is the subalgebra of diagonal matrices. These groups act on $\mathcal{M}$. The actions and their momentum mappings are given in the following formulas:

$$g \cdot \mathbf{m} = \begin{cases} \begin{cases} ((x^r)^T g_r^T, (u^r)^T g_r^{-1})_{r=1}^{\ell'}, \\ J_{G^{\ell'}} : \mathcal{M} \to \mathfrak{g}^{\ell'}, \quad J_{G^{\ell'}}(\mathbf{m}) = (x^r \otimes u^r)_{r=1}^{\ell'} \end{cases} \\ \begin{cases} (\mathbf{x}g^T, \mathbf{u}g^{-1}) = ((x^r)^T g^T, (u^r)^T g^{-1})_{r=1}^{\ell'}, \\ J_G : \mathcal{M} \to \mathfrak{g}, \quad J_G(\mathbf{m}) = \mathbf{x}^T \mathbf{u} = \sum_{r=1}^{\ell'} x^r \otimes u^r \end{cases} \\ \begin{cases} (g_\alpha x_\alpha, g_\alpha^{-1} u_\alpha)_{\alpha=1}^{\ell}, \\ J_{\mathbf{G}_\epsilon} : \mathcal{M} \to \mathfrak{g}_\epsilon, \quad J_{\mathbf{G}_\epsilon}(\mathbf{m}) = \operatorname{diag}(x_\alpha \cdot u_\alpha)_{\alpha=1}^{\ell}, \end{cases} \end{cases}$$

where in the first formula $g = (g_1, \ldots, g_{\ell'}) \in \mathbf{G}^{\ell'}$, in the second formula $g \in \mathbf{G}$ and in the third formula $g \in \mathbf{G}_\epsilon$.

(1.7) An Infinitesimal $\mathfrak{P}$ Action: Evaluation at the Eigenvalues of a. The Lie algebra $\mathfrak{P}$ acts on $\mathcal{M}$. The action and its momentum mapping are given for $\xi(\zeta) \in \mathfrak{P}$ by this formula:

$$\xi(\zeta) \cdot \mathbf{m} = (\mathbf{x}\xi(\zeta)^T, -\mathbf{u}\xi(\zeta))\rfloor_{\zeta=\mathbf{a}} = ((x^r)^T \xi(a_r)^T, -(u^r)^T \xi(a_r)),$$

$$\Upsilon : \mathcal{M} \to \mathfrak{Q}^0, \quad \Upsilon(\mathbf{m}) = \pi_{\mathfrak{Q}^0}(\mathbf{x}^T(\zeta - \mathbf{a})^{-1}\mathbf{u}) = \sum_{r=1}^{\ell'} \frac{x^r \otimes u^r}{\zeta - a_r}$$

where, for $\xi_j \in \mathfrak{g}$,

$$\mathbf{x}\xi_j^T \zeta^j \rfloor_{\zeta=\mathbf{a}} \doteq \mathbf{a}^j \mathbf{x}\xi_j^T.$$

The momentum map Υ is equivariant with respect to the infinitesimal coadjoint action of $\mathfrak{P}$ on $\mathfrak{P}^*$:

$$\begin{array}{ccc} \mathcal{M} & \xrightarrow{\Upsilon} & \mathfrak{P}^* \cong \mathfrak{Q}^0 = \mathfrak{Q} \\ {\scriptstyle \xi(\zeta)}\downarrow & & \downarrow{\scriptstyle -\pi_{\mathfrak{Q}^0} \circ \mathrm{ad}_{\xi(\zeta)}} \\ \mathcal{M} & \xrightarrow[\Upsilon]{} & \mathfrak{P}^* \cong \mathfrak{Q}^0 = \mathfrak{Q}. \end{array}$$

(1.8) Hamiltonian $\mathbf{H} \doteq \mathbf{GL}(\ell',\mathbb{C})$ Actions. Let $\mathbf{H}^\ell$ denote the ℓ fold direct product of $\mathbf{H}$ with itself. Then $\mathbf{H}^\ell$ is a Lie group whose Lie algebra may be identified with the ℓ fold direct sum of $\mathfrak{h}$ with itself. Let $\mathbf{H}_a$ be the isotropy subgroup of $\mathbf{a}$. It is the group of invertible diagonal matrices and its Lie algebra $\mathfrak{h}_a$ is the subalgebra of diagonal matrices. These groups act on $\mathcal{M}$. The actions and their momentum mappings are given in the following formulas:

$$h \cdot \mathbf{m} = \begin{cases} \begin{cases} (h_\alpha x_\alpha, (h_\alpha^{-1})^T u_\alpha)_{\alpha=1}^\ell, \\ J_{H^\ell} : \mathcal{M} \to \mathfrak{h}^\ell, \quad J_{H^\ell}(\mathbf{m}) = (x_\alpha \otimes u_\alpha)_{\alpha=1}^\ell \end{cases} \\ \begin{cases} (h\mathbf{x}, (h^{-1})^T \mathbf{u}) = (h x_\alpha, (h^{-1})^T u_\alpha)_{\alpha=1}^\ell, \\ J_H : \mathcal{M} \to \mathfrak{h}, \quad J_H(\mathbf{m}) = \mathbf{x}\mathbf{u}^T = \sum_{\alpha=1}^\ell x_\alpha \otimes u_\alpha \end{cases} \\ \begin{cases} (h_r x^r, h_r^{-1} u^r)_{r=1}^{\ell'}, \\ J_{H_a} : \mathcal{M} \to \mathfrak{h}_a, \quad J_{H_a}(\mathbf{m}) = \mathrm{diag}(x^r \cdot u^r)_{r=1}^{\ell'} \end{cases} \end{cases}$$

where in the first formula $h = (h_1, \ldots, h_\ell) \in \mathbf{H}^\ell$, in the second formula $h \in \mathbf{H}$ and in the third formula $h \in \mathbf{H}_a$.

(1.9) An Infinitesimal $\mathfrak{P}'$ Action: Evaluation at the Eigenvalues of ϵ. There are two interesting actions of $\mathfrak{P}'$ on $\mathcal{M}$. These action and their momentum mappings are given by this formula:

$$\eta(z) \cdot \mathbf{m} = \begin{cases} (\mathbf{x}\eta(0)^T, -\mathbf{u}\eta(0)), \quad \Sigma : \mathcal{M} \to \mathfrak{Q}', \quad \Sigma(\mathbf{m}) = z^{-1}\mathbf{x}\mathbf{u}^T \\ \\ (\eta(z)\mathbf{x}, -\eta(z)^T\mathbf{u})\lfloor_{z=\epsilon} = (\eta(\epsilon_\alpha)x_\alpha, -\eta(\epsilon_\alpha)^T u_\alpha), \\ \\ \Sigma : \mathcal{M} \to \mathfrak{Q}', \quad \Sigma(\mathbf{m}) = x(z-\epsilon)^{-1}u^T = \sum_{\alpha=1}^\ell \frac{x_\alpha \otimes u_\alpha}{z-\epsilon_\alpha} \end{cases}$$

where $\eta(z) \in \mathfrak{P}'$ and for $\eta_j \in \mathfrak{h}$,

$$\eta_j z^j \mathbf{x}\lfloor_{z=\epsilon} \doteq \eta_j \mathbf{x}\epsilon^j.$$

The momentum map Σ is equivariant with respect to the infinitesimal coadjoint action of $\mathfrak{P}'$ on $\mathfrak{P}^* \cong \mathfrak{Q}'$:

$$\begin{array}{ccc} \mathcal{M} & \xrightarrow{\Sigma} & \mathfrak{P}'^* \cong \mathfrak{Q}' \\ {\scriptstyle \eta(z)}\downarrow & & \downarrow{\scriptstyle -\pi_{\mathfrak{Q}'} \circ \mathrm{ad}_{\eta(z)}} \\ \mathcal{M} & \xrightarrow[\Sigma]{} & \mathfrak{P}'^* \cong \mathfrak{Q}'. \end{array}$$

An Involution Theorem. Let S^ϵ (S_a) denote translation by ϵ (**a**). Let

$$\mathcal{G}^\epsilon = \{\phi \circ S^\epsilon \circ \Upsilon\} \quad \text{and} \quad \mathcal{H}_a \doteq \{\phi \circ S_a \circ \Sigma\} \tag{1.10a}$$

over all ad* invariant functions ϕ on $L(\mathfrak{g})$ (respectively $L(\mathfrak{h})$) restricted to the Poisson manifold $\epsilon + \mathfrak{Q}$ (respectively $\mathbf{a} + \mathfrak{Q}'$). By the *AKS* theorem and the *GS* theorem, $\mathcal{G}^\epsilon$ is an involutive ring of functions on $\mathcal{M}$. Let

$$\begin{cases} \mathbb{L}(\mathbf{m}, z) = S_a \circ \Sigma(\mathbf{m}) = \mathbf{a} + \Sigma(\mathbf{m}) = \mathbf{a} + \mathbf{x}^T(z-\epsilon)^{-1}\mathbf{u} \\ \quad \text{and} \\ Z(\mathbf{m}, \zeta) = S^\epsilon \circ \Upsilon(\mathbf{m}) = \epsilon + \Upsilon(\mathbf{m}, \zeta) = \epsilon + \mathbf{x}(\zeta - \mathbf{a})^{-1}\mathbf{u}^T \end{cases} \tag{1.10b}$$

The coefficients in the characteristic polynomials $|\mathbb{L}(\mathbf{m}, z) - \zeta I_\ell|$ and $|Z(\mathbf{m}, \zeta) - zI_\ell|$, thought of as series in ζ and z, belong to $\mathcal{G}^\epsilon$ (respectively $\mathcal{H}_a$). The Weinstein/Aronczan formula [14],

$$|\mathbb{L}(\mathbf{m}, z) - \zeta I_{\ell'}| \cdot |\epsilon - zI_\ell| = |Z(\mathbf{m}, \zeta) - zI_\ell||\mathbf{a} - \zeta I_{\ell'}| \tag{1.10c}$$

implies that these coefficients belong to $\mathcal{G}^\epsilon \cap \mathcal{H}_a$.

Injective Moment Maps. The **H** action commutes with the **G** action and the $\mathbf{H}_a$ action commutes with the $\mathbf{G}^\ell$ action. Therefore J_H is constant on **G** orbits and $J_{\mathbf{G}^\ell}$ is constant on $\mathbf{H}_a$ orbits. Υ is also $\mathbf{H}_a$ invariant: $\mathbf{H}_a \cdot \mathbf{m} \subseteq \Upsilon^{-1}(\Upsilon(\mathbf{m})) \quad \forall \mathbf{m} \in \mathcal{M}$.

The $\mathbf{H}_a$ action preserves $\mathcal{M}^1$. Its restriction to $\mathcal{M}^1$ is free and proper; hence $\mathcal{M}^1/\mathbf{H}_a$ is a manifold [2]. It is a Poisson manifold under the identification of $C^\infty(\mathcal{M}^1/\mathbf{H}_a)$ with the $\mathbf{H}_a$ invariant functions on $\mathcal{M}^1$. Its symplectic leaves are of the form $J_{\mathbf{H}_a}^{-1}(\mathcal{O}_c)/\mathbf{H}_a$ where $\mathbf{c} \in \mathfrak{h}_a^* \cong \mathfrak{h}_a$ and $\mathcal{O}_c$ is its coadjoint orbit. These spaces are isomorphic to the *Marsden/Weinstein* reduced spaces $J_{\mathbf{H}_a}^{-1}(\mathbf{c})/\mathbf{H}_a$. Let us fix a point $\mathbf{m} \in \mathcal{M}^1$ and let

$$\mathcal{P}_c \doteq J_{\mathbf{H}_a}^{-1}(\mathbf{c})/\mathbf{H}_a \quad \text{where} \quad \mathbf{c} \doteq J_{\mathbf{H}_a}(\mathbf{m}) \tag{1.11a}$$

denote the symplectic leaf in $\mathcal{M}^1/\mathbf{H}_a$ through the $\mathbf{H}_a$ orbit of **m**. The $\mathbf{G}^\ell$ action preserves $\mathcal{M}^1$ and, since the $\mathbf{G}^\ell$ and $\mathbf{H}_a$ actions commute, there is an induced $\mathbf{G}^\ell$ action on $\mathcal{M}^1/\mathbf{H}_a$. The leaf $\mathcal{P}_c$ may be identified with the orbit $\mathbf{G}^\ell \cdot \mathbf{H}_a\mathbf{m}$:

$$\mathcal{P}_c \cong \mathbf{G}^\ell \cdot \mathbf{m}, \quad g \cdot \mathbf{H}_a\mathbf{m} \to \mathbf{H}_a g \cdot \mathbf{m}. \tag{1.11b}$$

In other words, the collection of reduced phase spaces $\mathcal{P}_c$ is parameterized by the points of $(\mathcal{M}^1/\mathbf{H}_a/\mathbf{G})^\ell$ which is isomorphic to $\mathcal{M}^1/(\mathbf{H}_a \times \mathbf{G}^\ell)$. Let $\mathfrak{Q}^1 \doteq \{\sum_{r=1}^{\ell'} \frac{\eta_r}{\zeta - a_r} \in \mathfrak{Q} : \eta_r \in \mathfrak{g},\ \text{rank}(\eta_r) = 1\}$. Then $\mathfrak{Q}^1$ is the image of $\mathcal{M}^1$ under Υ. If $\mathbf{m} \in \mathcal{M}^1$ then the $\mathbf{H}_a$ orbit of $\mathbf{m}$ is given by this formula:

(1.12a) $$\mathbf{H}_a \cdot \mathbf{m} = \Upsilon^{-1}(\nu) = J_{\mathbf{G}^\ell}^{-1}(\nu) \quad \text{where} \quad \nu = \Upsilon(\mathbf{m}).$$

See ([3], Proposition 2.4) for a proof. Therefore the mapping

(1.12b) $$\tilde{\Upsilon} : \mathcal{M}^1/\mathbf{H}_a \to \mathfrak{Q}^1, \quad \tilde{\Upsilon}(\mathbf{H}_a \cdot \mathbf{m}) = \Upsilon(\mathbf{m}),$$

is injective. It is a bracket preserving diffeomorphism of $\mathcal{M}^1/\mathbf{H}_a$ onto the finite dimensional space $\mathfrak{Q}^1$ ([3], Proposition 4.3). The restriction of $\tilde{\Upsilon}$ to a symplectic leaf $\mathcal{P}_c$ is a canonical mapping onto a symplectic leaf in $\mathfrak{Q}$. The symplectic leafs in $\mathfrak{P}^* \cong \mathfrak{Q}$ are coadjoint orbits. Thus $\mathcal{P}_c \cong \mathcal{O}_{\Upsilon(\mathbf{m})}$ where the coadjoint orbit is given by this formula:

(1.12c) $$\mathcal{P}_c \cong \mathcal{O}_{\Upsilon(\mathbf{m})}$$

and $\mathcal{O}_{\Upsilon(\mathbf{m})} =$

$$\left\{ \pi_{\mathfrak{Q}} \left(\mathrm{Ad}_{\exp(t\xi(\zeta))} \Upsilon(\mathbf{m}) \right) = \sum_{r=1}^{\ell'} \frac{\mathrm{Ad}_{\exp(t\xi(a_r))}\, x^r \otimes u^r}{\zeta - a_r} \ :\ \xi(\zeta) \in \mathfrak{P} \right\}.$$

The $\mathfrak{P}$ action does not preserve $\mathcal{M}^1$; nevertheless, one may consider an induced local $\mathfrak{P}$ action on $\mathcal{M}^1$. The moment map for the induced $\mathfrak{P}$ action on $\mathcal{M}^1/\mathbf{H}_a$ is $\tilde{\Upsilon}$.

There is an induced $\mathbf{G}$ action on $\mathcal{P}_c$ and $\mathbf{G}$ acts in a natural way on $\mathcal{O}_{\Upsilon(\mathbf{m})} \subseteq \mathfrak{Q}$; namely, $g \cdot \xi(\zeta) = g\xi(\zeta)g^{-1}$. The momentum mappings are given by these formulas:

(1.13a) $$\begin{array}{ccccc} \mathcal{P}_c & \xrightarrow{J_{G,c}} & \mathfrak{g} & J_{G,c}(\mathbf{m}) = J_G(\mathbf{m}) = \mathbf{x}^T\mathbf{u} & \\ \tilde{\Upsilon}\downarrow & & & \downarrow I & \\ \mathcal{O}_{\Upsilon(\mathbf{m})} & \xrightarrow[J]{} & \mathfrak{g} & J(\sum_{j=1}^{\infty} \xi_j \zeta^{-j}) = \xi_1. & \end{array}$$

These mappings are related by the following formula:

$$J_{G,c} = J \circ \tilde{\Upsilon}.$$

Let $\mathcal{P}_{c,\nu} \doteq J_{G,c}^{-1}(\nu)/\mathbf{G}_\nu$ be the reduced phase space containing $H_a \cdot \mathbf{m}$ where $\nu = \mathbf{x}^T\mathbf{u}$ and $\mathbf{G}_\nu$ is the isotropy subgroup in $\mathbf{G}$ of ν. The injective moment mapping induces an isomophism of the corresponding reduced spaces:

$$\text{(1.13b)} \qquad \mathcal{P}_{c,\nu} = J_{G,c}^{-1}(\nu)/\mathbf{G}_\nu \cong J^{-1}(\nu)/\mathbf{G}_\nu.$$

There is an induced $\mathbf{G}_\epsilon$ action on $\mathcal{P}_c$ and $\mathbf{G}_\epsilon$ acts in a natural way on $\mathcal{O}_{\Upsilon(\mathbf{m})} \subseteq \mathfrak{Q}$; namely, $g \cdot \xi(\zeta) = g\xi(\zeta)g^{-1}$. The momentum mappings are given by these formulas:

$$\text{(1.14a)} \qquad \begin{array}{ccccc} \mathcal{P}_c & \xrightarrow{J_{G,c,\epsilon}} & \mathfrak{g}_\epsilon & J_{G,c,\epsilon}(\mathbf{m}) = \operatorname{diag}(\mathbf{x}^T\mathbf{u}) \\ \tilde{\Upsilon}\big\downarrow & & & \big\downarrow I \\ \mathcal{O}_{\Upsilon(\mathbf{m})} & \xrightarrow[J_\epsilon]{} & \mathfrak{g}_\epsilon & J_\epsilon(\sum_{j=1}^{\infty} \xi_j \zeta^{-j}) = \operatorname{diag}(\xi_1). \end{array}$$

These mappings are related by this formula:

$$J_{G,c,\epsilon} = J_\epsilon \circ \tilde{\Upsilon}.$$

Let $\mathcal{P}_{c,\epsilon,\nu} \doteq J_{G,c,\epsilon}^{-1}(\nu)/\mathbf{G}_{\epsilon,\nu}$ be the reduced phase space containing $H_a \cdot \mathbf{m}$ where $\nu = \operatorname{diag}(\mathbf{x}^T\mathbf{u})$ and $\mathbf{G}_{\epsilon,\nu} = \mathbf{G}_\epsilon$ is the isotropy subgroup in $\mathbf{G}_\epsilon$ of ν. The injective moment mapping induces an isomophism of the corresponding reduced spaces:

$$\text{(1.14b)} \qquad \mathcal{P}_{c,\epsilon,\nu} = J_{G,c,\epsilon}^{-1}(\nu)/\mathbf{G}_\epsilon \cong J_\epsilon^{-1}(\nu)/\mathbf{G}_\epsilon.$$

Differential equations on our reduced spaces correspond to matrix equations under the injective momentum mappings. The Hamiltonian system corresponding to a function in $\mathcal{G}^\epsilon$ corresponds to a matrix Lax equation. The equations are given in [3], propositions (3.4) and (3.7).

Involutive Rings of Functions. The natural projection $\mathcal{M}^1 \to \mathcal{M}^1/\mathbf{H}_a$ is a Poisson map; hence, Υ being $\mathbf{H}_a$ invariant, $\mathcal{G}^\epsilon$ drops to an involutive ring on $\mathcal{M}^1/\mathbf{H}_a$. The natural projections

$$\text{(1.15a)} \qquad \mathcal{P}_c \to \begin{cases} \mathcal{P}_{c,\nu} \\ \text{or} \\ \mathcal{P}_{c,\epsilon,\nu} \end{cases}$$

are likewise Poisson mappings. Let $\mathcal{G}^0$ be defined in accordance with $\mathcal{G}^\epsilon$. By equivariance, $\Upsilon \circ g = g \circ \Upsilon$, any function in $\mathcal{G}^0$ is $\mathbf{G}$ invariant. Therefore $\mathcal{G}^0$ drops to an involutive ring of functions on $\mathcal{P}_{c,\nu}$. Also, by equivariance, any function $f = \phi \circ S^\epsilon \circ \Upsilon$ in $\mathcal{G}^\epsilon$ is $\mathbf{G}_\epsilon$ invariant:

$$\text{(1.15b)}\quad f(g \cdot \mathbf{m}) = \phi(\epsilon + g \cdot \Upsilon(\mathbf{m})) = \phi(g \cdot (\epsilon + \Upsilon(\mathbf{m}))) = \phi(\epsilon + \Upsilon(\mathbf{m})) = f(\mathbf{m}),$$

if $g \in \mathbf{G}_\epsilon$. Therefore $\mathcal{G}^\epsilon$ drops to an involutive ring of functions on $\mathcal{P}_{c,\epsilon,\nu}$. The characteristic polynomial of any matrix in $\mathfrak{Q}^1$ belongs to $\mathcal{G}^\epsilon$. Any matrix in $\mathfrak{Q}^1$ is isospectral with respect to the Hamiltonian flow generated by an element of $\mathcal{G}^\epsilon$.

Another Injective Moment Map and More Involutive Rings of Functions. Let

$$\text{(1.16a)}\quad \begin{cases} \mathfrak{h}_\ell \doteq S_a \circ J_H(\mathcal{M}_\ell) = \{\mathbf{a} + \mathbf{x}\mathbf{u}^T \mid (\mathbf{x}, \mathbf{u}) \in \mathcal{M}_\ell\} \\ \textit{and} \\ \mathfrak{h}_\ell^\epsilon \doteq \{\mathbf{a} + \mathbf{x}(I - \epsilon)^{-1}\mathbf{u}^T \mid (\mathbf{x}, \mathbf{u}) \in \mathcal{M}_\ell\}. \end{cases}$$

Then, in view of ([3]page 460),

$$\mathbf{G} \cdot \mathbf{m} = J_H^{-1}(J_H(\mathbf{m})) \implies$$

$$\text{(1.16b)}\quad \begin{cases} \mathcal{M}_\ell/\mathbf{G} \cong \mathfrak{h}_\ell & \text{under } S_a \circ J_H \\ \text{and the induced momentum mapping,} & \\ \tilde{J}_H : \mathcal{M}_\ell/\mathbf{G} \to \mathfrak{h}, \quad \tilde{J}_H(\mathbf{G} \cdot \mathbf{m}) = J_H(\mathbf{m}), & \text{is injective.} \end{cases}$$

The natural projection $\mathcal{M}_\ell \to \mathcal{M}_\ell/\mathbf{G}$ being a Poisson mapping, $\mathcal{G}^0$ drops to a Poisson commuting family of functions on $\mathcal{M}_\ell/\mathbf{G} \cong \mathfrak{h}_\ell$. The characteristic polynomial of any matrix in $\mathfrak{h}_\ell$ belongs to $\mathcal{G}^0$. Any matrix in $\mathfrak{h}_\ell$ is isospectral with respect to the Hamiltonian flow generated by an element of $\mathcal{G}^0$. By Υ equivariance, $\mathcal{G}^\epsilon$ drops to a Poisson commuting family of functions on $\mathcal{M}_\ell/\mathbf{G}_\epsilon$. The characteristic polynomial of any matrix in $\mathfrak{h}_\ell^\epsilon$ belongs to $\mathcal{G}^\epsilon$. Any matrix in $\mathfrak{h}_\ell^\epsilon$ is isospectral with respect to the Hamiltonian flow generated by an element of $\mathcal{G}^\epsilon$.

CHAPTER II

A NEUMANN SYSTEM FOR THE AKNS PROBLEM

A Derivation of the Equations of Motion

Riemann Surface and Meromoprphic Functions. Let $\mathcal{R}$ be a Riemann Surface and let $g_{\mathcal{R}}$ be its genus. Choose ℓ distinct points ∞_α of $\mathcal{R}$ and let $\infty = \infty_1 + \cdots + \infty_\ell$. Suppose there exists a meromorphic function ζ with a simple pole at each ∞_α and no other poles. Then ∞ is the polar divisor of ζ and ζ^{-1} is a local parameter at each ∞_α. Let $\mathfrak{A}_\infty$ be the ring of all meromorphic functions on $\mathcal{R}$ holomorphic in $\mathcal{R} \setminus \infty$.

A Weierstrass gap number of ∞ is an integer $w = lq + \alpha$, with $(\alpha = 0, \ldots, \ell - 1)$ and $q \geq 0$, such that $q\infty + \infty_1 + \cdots + \infty_\alpha$ is not the divisor of a function in $\mathfrak{A}_\infty$. Let W denote the set of Weierstrass gap numbers of ∞. It is a known that W consists of $g_{\mathcal{R}}$ positive integer. Our function ζ belongs to $\mathfrak{A}$ and therefore W does not contain any multiples of ℓ.

Suppose, in addition, there is given a meromorphic function z on $\mathcal{R}$ satisfying the following 3 conditions.

(2.1a) Let $(z)_\infty = (1) + \cdots + (\ell')$, $(r) \in \mathcal{R}$, denote the polar divisor of z and let $a_r = \zeta(r)$. It shall be assumed that the a_r are distinct. It follows that each (r) is a simple pole of z and z^{-1} may serve as a local parameter at (r). Let **a** be the matrix $\operatorname{diag}(a_1, \ldots, a_{\ell'})$.

(2.1b) It is assumed that the order of z at ∞_α is nonnegative and independent of α. Let m be that order. Then z is given in a neighborhood of ∞_α by a formula like this one:

$$z = \zeta^{-m}(\epsilon_\alpha + \sum_{j=1}^{\infty} z_{\alpha,j}\zeta^{-j})$$

for some constants ϵ_α and $z_{\alpha,j}$. Let

$$\epsilon = \operatorname{diag}(\epsilon_1, \ldots, \epsilon_\ell) \quad \text{and} \quad z_j = \operatorname{diag}(z_{1,j}, \ldots, z_{\ell,j}).$$

(2.1c) The genus $g_{\mathcal{R}}$ of $\mathcal{R}$ is related to the parameters m, ℓ and ℓ' by this formula:

$$g_{\mathcal{R}} = (\ell - 1)(\ell' - 1) - \frac{1}{2}\ell(\ell - 1)m.$$

The last condition does not enter directly into the construction of this section. Its significance lies in the role of the Jacobian of $\mathcal{R}$ as an level surface of the constants of motion found below. All this will be clarified in Chapter IV.

Divisors. Let δ be positive divisor of degree $g_{\mathcal{R}} + \ell - 1$ with support in $\mathcal{R} \setminus \infty$ and satisfying this condition:

$$L(\delta - \infty) = \{0\}. \tag{2.2a}$$

Then any function with poles no worse than δ and vanishing on ∞ is identically 0. It follows then that $L(\delta)$ is spanned by ℓ functions $f_1, \cdots, f_\ell$ such that $f_\alpha(\infty_\beta) = \delta_{\alpha,\beta}$. The condition (2.2a) implies that δ is a nonspecial divisor. Conversely, any nonspecial divisor of degree $g_{\mathcal{R}} + \ell - 1$ with support in $\mathcal{R} \setminus \infty$ satisfies (2.2a). Let

$$\delta_\alpha = \delta + (f_\alpha) - \infty + \infty_\alpha = (\delta - (f_\alpha)_\infty) + ((f_\alpha)_0 - (\infty - \infty_\alpha)). \tag{2.2b}$$

Then each δ_α is an integral divisor of degree $g_{\mathcal{R}}$ satisfying this analogue of (2.2a):

$$L(\delta_\alpha - \infty_\alpha) = \{0\}. \tag{2.2c}$$

For if $f \in L(\delta_\alpha - \infty_\alpha)$ then $ff_\alpha \in L(\delta - \infty) = \{0\}$ and therefore $f = 0$.

Choose a basis $\boldsymbol{\omega} = (\omega_i)_{i=1}^{g_{\mathcal{R}}}$ for $\Gamma(\mathcal{R}, \Omega^1)$ and $2g_{\mathcal{R}}$ simple closed curves $\mathbf{a} = (a_i)_{i=1}^{g_{\mathcal{R}}}, \mathbf{b} = (b_i)_{i=1}^{g_{\mathcal{R}}}$ spanning $H_1(\mathcal{R}, \mathbb{Z})$ with the standard intersection matrix. More details about the basis will not be needed. Let Γ denote the lattice in $\mathbb{C}^{g_{\mathcal{R}}}$ generated by the columns of the $g_{\mathcal{R}} \times 2g_{\mathcal{R}}$ matrix $\int_{(\mathbf{a},\mathbf{b})} \boldsymbol{\omega}$. The Jacobian of $\mathcal{R}$, $\mathcal{J}(\mathcal{R})$, is the complex Abelian group $\mathbb{C}^g_{\mathcal{R}}/\Gamma$. Let $\mathcal{A}$ be the Abel mapping,

$$\mathcal{A} : \text{Divisors mod Linear Equivalence} \longrightarrow \mathcal{J}(\mathcal{R}) : \quad \mathcal{A}(\delta) = \int_{p_0}^{\delta} \omega.$$

Then δ_α is the unique integral divisor of degree $g_{\mathcal{R}}$ such that

$$\mathcal{A}(\delta - (\infty - \infty_\alpha)) = \mathcal{A}(\delta_\alpha).$$

The set of all integral divisors of degree $g_{\mathcal{R}}$ satisfying (2.2b) is isomorphic to the affine open subset $\mathcal{J}(\mathcal{R}) \setminus \Theta$ under the Abel mapping; Θ denotes the zero locus of the corresponding Riemann theta function. Thus the set of all integral divisors δ satisfying (2.2a) is isomorphic to $\mathcal{J}(\mathcal{R}) \setminus \Theta$.

(2.3) PROPOSITION: Let $\mathcal{T}$ denote translation by the zero divisor of z less ∞:

$$\mathcal{T} : \mathrm{Div}(\mathcal{R}) \to \mathrm{Div}(\mathcal{R}) : \quad \mathcal{T}(\delta) = \delta + (z)_0 - \infty.$$

(2.3a) It is, up to linear equivalence, an isomorphism of the set of positive divisors of degree $g_{\mathcal{R}} + \ell - 1$ into the set of positive divisors of degree $g_{\mathcal{R}} + \ell' - 1$.

(2.3b) If δ satisfies (2.2a) then its image Δ satisfies this analogous condition:

(2.3c)
$$L(\Delta - (z)_\infty) = \{0\}.$$

Proof: This mapping is well defined modulo linear equivalence and its push forward under the Abel map is a translation in the group $\mathcal{J}(\mathcal{R})$. The mapping $\mathcal{T}$, modulo linear equivalence, is an isomorphism because translation in a topological group is an isomorphism. To explain this let δ be a positive divisor of degree $g_{\mathcal{R}} + \ell - 1$. Then $\Delta \doteq \delta + (z)_0 - \infty$ is a divisor of degree $g_{\mathcal{R}} + \ell' - 1$. By the Riemann-Roch theorem $\dim(L(\Delta)) \geq \ell'$. If Δ' is an integral divisor linearly equivalent to Δ then $\Delta' = (f) + \Delta$ for some meromorphic function f. Now $f \in L(\Delta)$ and the preimage of Δ' under $\mathcal{T}$ is linearly equivalent to δ. This proves (a). The proof of (b) hinges on the following fact. If f belongs to $L(\delta - (z)_\infty)$ then zf belongs to $L(\Delta - \infty)$. $\square$

The proof of the following proposition is based on the Riemann-Roch theorem. See [18a] for details.

(2.4) PROPOSITION: Let δ be a positive divisor of degree $g_{\mathcal{R}} + \ell - 1$ satisfying (2.2a). Then there exists a unique Abelian differential Ω such that

(2.4a)
$$(\Omega) - \delta + 2\infty \geq 0$$

and

(2.4b)
$$\Omega = -\zeta^2(1 + O(\zeta^{-1}))d\zeta^{-1} \text{ at } \infty_\alpha$$

for each $(\alpha = 1, \ldots, \ell)$.

Since the degree of the divisor of any Abelian differential is $2g_{\mathcal{R}} - 2$, there exist a unique positive divisor δ^ι of degree $g_{\mathcal{R}} + \ell - 1$ such that

(2.4c)
$$(\Omega) = \delta + \delta^\iota - 2\infty.$$

The argument in the proof of (2.3c) may be used to show that δ^ι satisfies (2.2a). We shall refer to δ^ι as the *dual* of δ. It is clear that the mapping ι is an involution of the set of positive divisors of degree $g_{\mathcal{R}} + \ell - 1$.

Let δ be a divisor on $\mathcal{R} \setminus \infty$ of degree $g_{\mathcal{R}} + \ell - 1$ satisfying (2.2a). By proposition (2.4), there exists an Abelian differential Ω and a divisor δ^ι of degree $g_{\mathcal{R}}$ such that

$$\textbf{(2.5a)} \qquad (\Omega) = \delta + \delta^\iota - 2\infty \quad \text{and} \quad \Omega = -\zeta^2(1 + O(\zeta^{-1}))d\zeta^{-1} \text{ at } \infty_\alpha.$$

Let $\Delta = \delta + (z)_0 - \infty$. By proposition (2.4), with ∞ replaced by $(z)_\infty$, there exists an Abelian differential Λ and a divisor Δ^ι of degree $g_{\mathcal{R}} + \ell' - 1$ such that

$$\textbf{(2.5b)} \qquad (\Lambda) = \Delta + \Delta^\iota - 2\infty \quad \text{and} \quad \Lambda = -z^2(\rho_r + O(z^{-1}))dz^{-1} \text{ at } (r)$$

where $\rho_r = \operatorname{Res}_{(r)}(z\Omega)$. Proposition (2.4) may be used to show that

$$\textbf{(2.5c)} \qquad \Delta^\iota = \delta^\iota + (z)_0 - \infty \quad \text{and} \quad \Lambda = z^2\Omega.$$

We have allowed ourselves a slight abuse of notation by using ι for the mapping (2.5b) associated with $(z)_\infty$ and the mapping (2.5a) associated with ∞. With this in mind, one might say that ι commutes with translation by the divisor $(z)_0 - \infty$.

Baker Functions. Let δ be a positive divisor on $\mathcal{R} \setminus \infty$ of degree $g_{\mathcal{R}} + \ell - 1$ satisfying (2.2a). Define divisors δ_α of degree $g_{\mathcal{R}}$ as in the formula preceeding (2.2b). Let Δ be a positive divisor in $\mathcal{R} \setminus (z)_\infty$ of degree $g_{\mathcal{R}} + \ell' - 1$ satisfying (2.3c). Let δ^ι be the divisor dual to δ and let Δ^ι be the divisor dual to Δ. Let

$$\theta(t,p) = \sum\nolimits_W \zeta^k z^\alpha t_{k,\alpha}$$

where $\theta_{k,\alpha}$ is a complex parameters and the sum is taken over all integer pairs (k, α) such that $\operatorname{ord}_\infty(\zeta^k z^\alpha) = nk - m\alpha$ belongs to W. The sum covers all of W because m and n are relatively prime. The vector $t \doteq (\theta_{k,\alpha})$ belongs to $\mathbb{C}^{g_{\mathcal{R}}}$. Let

$$\partial_s = \partial_{k,\alpha} = \frac{\partial}{\partial t_{k,\alpha}} \qquad \text{if } s = nk - m\alpha.$$

Let τ be another complex parameter. Let $\partial = \frac{\partial}{\partial \tau} = \cdot$. Let $\mathcal{W}$ be the linear space consisting of all functions χ satisfying the following 2 conditions:

(2.6a) Any pole of χ lies in δ and **(2.6b)** $\chi\exp(-\epsilon_\alpha\zeta\tau-\theta)$ holomorphic at ∞_α.

Then $\mathcal{W}$ is ℓ dimensional for all t in some open set in $\mathbb{C}^{g_\mathcal{R}}$ containing 0. See [18a] for details. It contains a unique function $\psi_\alpha=\psi_{\delta,\alpha}(t,p)$ given at the pole ∞_β of ζ by this formula:

(2.6c) $$\psi_\alpha=(\delta_{\alpha,\beta}+O(\zeta^{-1}))\exp(\epsilon_\alpha\zeta\tau+\theta).$$

The vector valued function $\boldsymbol{\psi}\doteq(\psi_1,\ldots,\psi_\ell)^T$ is called the Baker function of δ. Let $\boldsymbol{\phi}(t,p)=\boldsymbol{\psi}^t_{\delta}(-t,p)$. Then $\boldsymbol{\phi}$ is called the Baker function dual to $\boldsymbol{\psi}$.

There is an interesting formula for the entries of ψ_α in terms of a scalar Baker function. Let us consider the linear space consisting of all functions χ satisfying the following 2 conditions:

(2.6d) Any pole of χ lies in δ_α and **(2.6e)** $\chi\exp(-\epsilon_\alpha\zeta\tau-\theta)$ holomorphic ∞_α.

This linear space is one dimensional for all (τ,t) in an open set containing 0. It contains a unique function χ_α given at the pole ∞_α of ζ by this formula:

(2.6f) $$\chi_\alpha=(1+O(\zeta^{-1}))\exp(\epsilon_\alpha\zeta\tau+\theta).$$

Then ψ_α is given by this formula:

(2.6g) $$\psi_\alpha=f_\alpha\chi_\alpha.$$

Let us consider the linear space consisting of all functions χ satisfying these 2 conditions:

(2.7a) Any pole of χ lies in Δ and **(2.7b)** $\chi\exp(\theta)$ is holomorphic at each (r).

Then this linear space is ℓ' dimensional for all t in some open set in $\mathbb{C}^{g_\mathcal{R}}$ containing 0. See [18a] for details. It contains a unique function $\Psi^r=\Psi^r_\Delta(t,p)$ given at the pole (s) of z by this formula:

(2.7c) $$\Psi^r=(\delta_{r,s}+O(z^{-1}))\exp(-\theta).$$

The vector $\Psi \doteq (\Psi^1, \ldots, \Psi^\ell)^T$ is called the Baker function of Δ. Let $\Phi(t,p) = \Psi^\iota_\Delta(-t,p)$. Then Φ is called the Baker function dual to Ψ.

The Baker functions satisfy certain linear differential equations. We shall describe these in terms of the operator formalism of Krichever [12]with the improvements by Cherednik [6]and Flaschka [9c].

Let ξ_j^α and η_j^α be the vectors in $\mathbb{C}^\ell$ defined in accordance with (2.6c) in a neighborhood of ∞_α by these formulas:

$$\begin{cases} \psi = \left(e_\alpha + \sum_{j=1}^\infty \xi_j^\alpha(\tau,t)\zeta^{-j}\right)\exp(\epsilon_\alpha\zeta\tau + \theta) \\ \text{and} \\ \phi = \left(e_\alpha + \sum_{j=1}^\infty \eta_j^\alpha(\tau,t)\zeta^{-j}\right)\exp(-\epsilon_\alpha\zeta\tau - \theta). \end{cases}$$

Let ξ_j be the $\ell \times \ell$ matrix whose column α is ξ_j^α. Let

$$\mathbf{z}_\infty = \zeta^{-m}\left(\epsilon + \sum_{j=1}^\infty z_1\zeta^{-j}\right), \quad \theta_\infty = \sum\nolimits_W \zeta^k \mathbf{z}_\infty^\alpha t_{k,\alpha},$$

$$\psi_\infty = \left(I + \sum_{j=1}^\infty \xi_j(\tau,t)\zeta^{-j}\right)\exp(\epsilon\zeta\tau + \theta_\infty),$$

and

$$\phi_\infty = \left(I + \sum_{j=1}^\infty \eta_j(\tau,t)\zeta^{-j}\right)\exp(-\epsilon\zeta\tau - \theta_\infty).$$

The τ derivative of ψ_∞ satisfies this formula:

$$(\dot\psi_\infty - (\epsilon\zeta + [\xi_1,\epsilon])\psi_\infty)\exp(-\epsilon\zeta\tau - \theta_\infty)$$

$$= \sum_{j=1}^\infty \left([\xi_{j+1},\epsilon] + \dot\xi_j - [\xi_1,\epsilon]\xi_j + \sum_{l=1}^j \xi_l z_{j+1-l}\right)\zeta^{-j}. \tag{2.8a}$$

This formula implies that $\dot\psi(\tau,t,p) - (\epsilon\zeta + [\xi_1,\epsilon])\psi(\tau,t,p)$ is $O(\zeta^{-1})$ at each ∞_α; hence,

$$\dot\psi(\tau,t,p) = B(\mathbf{q}(\tau,t),\zeta)\psi(\tau,t,p) \quad \text{where} \quad B \doteq \begin{cases} \epsilon\zeta + \mathbf{q} \\ \text{and} \\ \mathbf{q} = [\xi_1,\epsilon], \end{cases} \tag{2.8b}$$

because the components of ψ are a basis for the linear space (2.6). A similar calculation, together with the formula

$$0 = \sum_{p \in \mathcal{R}} \mathrm{Res}_p(\psi \otimes \phi\Omega) = \sum_{\alpha=1}^{\ell} \mathrm{Res}_{\infty_\alpha}(\psi \otimes \phi\Omega)$$

(2.8c)
$$= \sum_{\alpha=1}^{\ell} (e_\alpha \otimes \eta_1^\alpha + \xi_1^\alpha \otimes e_\alpha + \Omega_\alpha(e_\alpha \otimes \epsilon_\alpha)) = \xi_1^T + \eta_1 + \Omega_1,$$

where $\Omega_1 = \mathrm{diag}(\Omega_1^\alpha)$ and $\Omega = -\zeta^2(1+\Omega_1^\alpha+\cdots)d\zeta^{-1}$ at ∞_α, shows that ϕ satisfies the adjoint of the equation (2.8b),

(2.8d)
$$\dot{\phi}(\tau, t, p) = -B(\mathbf{q}(\tau, t), \zeta)^T \phi(\tau, t, p).$$

(2.9) LEMMA: There exists a unique monic differential operator $L_k = \epsilon^{-k}\,\partial^k$ plus lower order terms in ∂ such that

(2.9a)
$$L_k \psi_\infty \cdot \psi_\infty^{-1} = \zeta^k + O(\zeta^{-1})$$

Proof: It is clear that $L_0 = I$ satisfies (2.9a) and $\epsilon_{0,j} = 0$. Suppose $L_0, L_1, \ldots, L_k$ are known. Then, using (2.9a), there exists matrix valued functions $\epsilon_{k,j}$ such that

(2.9b)
$$L_k \psi_\infty \cdot \psi_\infty^{-1} = \zeta^k + \sum_{j=1}^{\infty} \epsilon_{k,j}\zeta^{-j}.$$

The series $\sigma \doteq \dot{\psi}_\infty \cdot \psi_\infty^{-1}$ has the following form:

(2.9c)
$$\sigma \doteq \dot{\psi}_\infty \cdot \psi_\infty^{-1} = \epsilon\zeta + \sum_{j=0}^{\infty} \sigma_j(\tau, t)\zeta^{-j},$$

for some functions σ_j. The τ derivative of (2.9b) is given by this formula:

$$\partial\, L_k \psi_\infty = \partial\, \psi_\infty \zeta^k + \left(\sum_{j=1}^{\infty} \epsilon_{k,j}\zeta^{-j}\right) \partial\, \psi_\infty + \left(\sum_{j=1}^{\infty} \dot{\epsilon}_{k,j}\zeta^{-j}\right) \psi_\infty$$

and therefore

$$\partial\, L_k \psi_\infty \cdot \psi_\infty^{-1} = \zeta^k \sigma + + \left(\sum_{j=1}^{\infty} \epsilon_{k,j}\zeta^{-j}\right) \sigma + \sum_{j=1}^{\infty} \dot{\epsilon}_{k,j}\zeta^{-j}.$$

This implies that the operator

(2.9d)
$$\epsilon L_{k+1} \doteq \partial\, L_k - \sum_{j=0}^{k} \sigma_j L_{k-j} - \epsilon_{k,1}\epsilon$$

satisfies (2.9a).

The L_k are unique for the following reason. Suppose there are 2 operators satisfying (2.9a). Let B be their difference and let s be the order of B. If $s \geq 0$ then $B\psi \cdot \psi^{-1}$ would go like a polynomial of degree s in ζ at ∞. This is a contradiction and therefore $B = 0$. □

Using lemma (2.9), the arguments leading to (2.8b) and the formula

$$\partial_{k,\alpha}\psi_\infty = \psi_\infty z_\infty^\alpha \zeta^k + O(\zeta^{-1})\exp(\epsilon\zeta\tau + \theta_\infty),$$

there exists a differential operator $L_{k,\alpha}$, which is a $\mathbb{C}$ linear combination of L_k such that

$$\textbf{(2.10a)} \qquad \partial_{k,\alpha}\psi = L_{k,\alpha}\psi.$$

Our next task is to show that the dual function satisfies the corresponding adjoint equation:

$$\textbf{(2.10b)} \qquad \partial_{k,\alpha}\phi = \epsilon^{-1}L_{k,\alpha}^\dagger\epsilon\phi.$$

(2.11) PROPOSITION: If $\eta(\zeta) \doteq -\zeta I + \sum_{k=0}^\infty \epsilon_{k,1}\zeta^{-k}$ then

$$\psi_\infty \Omega_\infty \phi_\infty^T = \zeta\,\eta(\zeta)\,d\zeta^{-1}.$$

Proof: Any pole of the Abelian differential $(L_k\psi) \otimes \phi\Omega$ lies in ∞. Let $\Omega_\infty = -\zeta^2(I + \sum_{j=1}^\infty \Omega_j\zeta^{-j})d\zeta^{-j}$ be the diagonal matrix whose entry α contains the Laurent series for Ω at ∞_α. Observe that the entry (i,j) of the series $\psi_\infty\Omega_\infty\phi_\infty^T$ is $\sum_{\alpha=1}^\ell \psi_{\infty,i,\alpha}\phi_{\infty,j,\alpha}\Omega_{\infty,\alpha,\alpha}$ and therefore

$$\mathrm{Res}_{\zeta=\infty}(\psi_\infty\Omega_\infty\phi_\infty^T) = \sum_{\alpha=1}^{\ell} \mathrm{Res}_{\infty_\alpha}(\psi \otimes \phi\Omega).$$

This proposition follows from the formula

$$(L_k\psi_\infty)\Omega_\infty\phi_\infty^T = L_k\psi_\infty \cdot \psi_\infty^{-1}(\psi_\infty\Omega_\infty\phi_\infty^T),$$

the residue formula $\mathrm{Res}_{\zeta=\infty}(L_k\psi_\infty \cdot \phi_\infty\Omega) = 0$ and (2.9b). □

(2.12) PROPOSITION: The dual Baker function satisfies the following analogue of the formula (2.9b):

(2.12a). $$\epsilon^{-1}L_k^{\dagger}\epsilon\phi_\infty \cdot \phi_\infty^{-1} = \zeta^k I + O(\zeta^{-1})$$

Proof: Set $M_0 = I$ and suppose $M_0, \ldots, M_k$ are known. Let $\epsilon^*_{k,j}$ be defined by this formula:

(2.12b) $$M_k\phi_\infty \cdot \phi_\infty^{-1} = \zeta^\kappa + \sum_{j=1}^{\infty} \epsilon^*_{k,j}\zeta^{-j}.$$

The M_k are determined by $M_0 = I$ and this analogue of (2.9d):

(2.12c) $$-\epsilon M_{k+1} = \partial M_k - \sum_{j=0}^{k} \rho_j M_{k-j} + \epsilon^*_{k,1}\epsilon \quad \text{where} \quad \rho = -\epsilon\zeta + \sum_{j=0}^{\infty} \rho_j \zeta^{-j}.$$

The formula $\epsilon^*_{k,1} = \epsilon^T_{k,1}$ follows from proposition (2.11). Let

$$L(\zeta) = \sum_{k=0}^{\infty} L_k\zeta^{-k} \quad \text{and} \quad M(\zeta) = \sum_{k=0}^{\infty} M_k\zeta^{-k}.$$

In this notation the formulas (2.9d) and (2.12c) are equivalent to these formulas:

(2.12d) $$(\partial - \sigma)L = \eta(\zeta)\epsilon \quad \text{and} \quad (\partial - \rho)M = -\eta(\zeta)^T\epsilon$$

The formula

(2.12e) $$(\partial - \sigma)\eta(\zeta) = \eta(\zeta)(\partial + \rho^T)$$

follows by differentiating (2.11) using the constancy of Ω. Then

$$(\partial - \sigma)\eta(\zeta) = \eta(\zeta)(\partial + \rho^T) = (\partial - \sigma)L\epsilon^{-1}(\partial + \rho^T)$$

implies

$$\eta(\zeta) = L\epsilon^{-1}(\partial + \rho^T) \quad \text{and} \quad \eta^T = (-\partial + \rho)\epsilon^{-1}L^{\dagger}.$$

When compared to the second formula in (2.12c), this proves

(2.12f) $$M = \epsilon^{-1}L^{\dagger}\epsilon.$$

and our proposition. □

(2.13a) REMARK: The proof of proposition (2.12) did not use the fact that ψ satisfies a first order equation; namely (2.8d). Under this assumption and in view of (2.8a), σ is very simple, $\sigma = \epsilon\zeta + \mathbf{q}$.

(2.13b) REMARK: The formula (2.12f) is illustrated by (2.8b) and (2.8d) where $L_1 = \epsilon^{-1}(\partial - \mathbf{q})$.

Soliton Hierarchy. The operator equations (2.10a) and (2.10b) are equivalent to a pair of matrix equations of the following form:

$$\text{(2.14)} \qquad \begin{cases} \partial_{k,\alpha}\psi(\tau,t,p) = B_{k,\alpha}(\mathbf{q}(\tau,t),\zeta(p))\psi(\tau,t,p) \\ \text{and} \\ \partial_{k,\alpha}\phi(\tau,t,p) = -B_{k,\alpha}(\mathbf{q}(\tau,t),\zeta(p))^T\phi(\tau,t,p). \end{cases}$$

where, as our notation suggests, each entry of $B_{k,\alpha}$ is a differential polynomial in the entries q. The integrability condition underlying our combined linear equations is a hierarchy of nonlinear partial differential equations disquised in their Lax representation. The $\ell = 2$ setup is an algebric version of the AKNS problem. The AKNS hierarchy includes the modified Korteweg de Vries equation and the nonlinear Schroedinger equation ([1] and [10]).

Trace Formulas and the AKNS/Neumann Hierarchy. Let $\mathbf{m}(\tau,t)$ be the point in $\mathbb{C}^{2\ell\ell'}$ defined by evaluating the Baker functions ψ and ϕ over the polar divisor of z:

$$x^r(\tau,t) = \rho_r\psi(\tau,t,r) \quad \text{and} \quad u^r(\tau,t) = \rho_r\phi(\tau,t,r)$$

where $\rho_r = \sqrt{\operatorname{Res}_r z\Omega}$. Let

$$x^r_\alpha = x_{r,\alpha} \quad \text{where} \quad \mathbf{x} = (x^1,\dots,x^{\ell'})^T = (x_1,\dots,x_\ell).$$

This is an $\ell' \times \ell$ matrix and x_α is a vector whose r^{th} component is α^{th} component of x^r. A matrix $\mathbf{u}$ is defined in the same way in terms of the u^r. Let $\mathbf{m} = (\mathbf{x},\mathbf{u})$.

(2.15) DEFINITION: Let M denote the subspace of $\mathbb{C}^{2n\ell}$ consisting of all points $\mathbf{m} \in \mathbb{C}^{2\ell\ell'}$ satisfying the **off diagonal** part of the following equations:

$$\text{(2.15a)} \qquad \sum_{r=1}^{\ell'} a^j_r x^r \otimes u^r = \mathbf{x}^T\mathbf{a}^j\mathbf{u} = \epsilon\delta_{j,m-1}.$$

where ($j = 0, \dots m-1$). These equations are called *constraints*. The $m\ell(\ell-1)$ constraints are independent and M is a smooth algebraic manifold of dimension

$$\text{(2.15b)} \qquad \dim(M) = 2\ell\ell' - m\ell(\ell-1) = 2g_R + 2(\ell+\ell'-1).$$

The second equation follows from (2.1c). If $m = 0$ then $M = \mathbb{C}^{2\ell\ell'}$.

The entries of the $\ell \times \ell$ matrix $(\psi \otimes \phi)\zeta^m z\Omega$, (m as in (2.1a)), are meromorphic differentials with polar divisor $(z)_\infty + 2\infty$. The residue formula applied to these differentials is equivalent to this one:

$$\text{(2.16a)} \qquad \mathbf{q} = \sum_{r=1}^{\ell'} a_r^m x^r \otimes u^r - z_1 = \mathbf{x}^T \mathbf{a}^m \mathbf{u} - z_1.$$

By (2.8b), $\mathbf{q}$ diagonal free. The diagonal part of (2.16a) is this:

$$\text{(2.16b)} \qquad z_1 = \operatorname{diag}(\mathbf{a}^m x_\alpha \cdot u_\alpha)_{\alpha=1}^{\ell}.$$

Thus the entries of $\mathbf{q}$ are functions of $\mathbf{m}$. We use the notation $\mathbf{q}(\mathbf{m})$ to indicate this dependence. The equations (2.8b) and (2.8d) with $p = (r)$ are equivalent to the following equations which define a vector field X on $\mathbb{C}^{2\ell\ell'}$, $\dot{\mathbf{m}} = X(\mathbf{m})$ or

$$\text{(2.17a)} \qquad \begin{cases} \dot{x}^r = (a_r\epsilon + \mathbf{q}(\mathbf{m}))x^r = (a_r\epsilon - z_1)x^r + \sum_{s=1}^{\ell'} a_s^m (x^r \cdot u^s) x^s \\ \\ \dot{u}^r = -(a_r\epsilon + \mathbf{q}(\mathbf{m})^T)u^r = -(a_r\epsilon - z_1)u^r - \sum_{s=1}^{\ell'} a_s^m (x^s \cdot u^r) u^s. \end{cases}$$

Let $\kappa'(\mathbf{m}) = \mathbf{x}\mathbf{u}^T = \sum_{\alpha=1}^{\ell} x_\alpha \otimes u_\alpha$. Then the equations (2.17a) are equivalent to these:

$$\text{(2.17b)} \qquad \begin{cases} \dot{x}_\alpha = (e_\alpha - z_{\alpha,1} + \kappa'(\mathbf{m}))x_\alpha = \epsilon_\alpha \mathbf{a} x_\alpha + \sum_{\beta\neq\alpha} (\mathbf{a}^m x_\alpha \cdot u_\beta) x_\beta \\ \\ \dot{u}_\alpha = -(e_\alpha - z_{\alpha,1} + \kappa'(\mathbf{m}))u_\alpha = -\epsilon_\alpha \mathbf{a} u_\alpha - \sum_{\beta\neq\alpha} (\mathbf{a}^m x_\beta \cdot u_\alpha) u_\beta \end{cases}$$

The nonlinear system (2.17a); equivalently (2.17b), are the object of our analysis. Its relationship to the soliton hierarchy of the AKNS problem is analogous to the well known relationship between the Neumann system and the Korteweg-de Vries hierarchy.

The equations (2.14) with $p = (r)$ define a vector field on $\mathbb{C}^{2\ell\ell'}$ which we call $X_{k,\alpha}$. The equations are given by these formulas:

$$\text{(2.18)} \qquad \partial_{k,\alpha} \mathbf{m} = X_{k,\alpha}(\mathbf{m}) : \quad \begin{cases} \partial_{k,\alpha} x^r = B_{k,\alpha}(\mathbf{q}(\mathbf{m}(t)), a_r) x^r \\ \\ \partial_{k,\alpha} u^r = -B_{k,\alpha}(\mathbf{q}(\mathbf{m}(t)), a_r)^T u^r. \end{cases}$$

The equations are autonomous because, according to Cherednik [6], the coefficients of the operators L_j are functions of the entries of $\mathbf{q}$. The collection of all such vector fields forms what we call the AKNS/Neumann hierarchy.

(2.19) PROPOSITION: Each curve $t_{k,\alpha} \to \mathbf{m}(\tau, t)$ lies on M.

This follows directly from the residue formula,

$$\sum_{p \in \mathcal{R}} \operatorname{Res}_p(\psi \otimes \phi)\zeta^j z\Omega = 0$$

where $(j = 0, \ldots, m-1)$.

Isospectral Matrices

The following technical but mild assumption is necessary if we wish to continue with our construction. We shall assume that $\mathfrak{A}_\infty$ separates points in the following way: for each pole (r) of z, there exists a y in $\mathfrak{A}_\infty$ that assumes distinct values at the points of $\mathcal{R}$ over α_r. According to ([18b], formula (1.3.8)), this assumption implies that the functions $\psi(\tau, t, p) \cdot u^r$ and $x^r \cdot \phi(\tau, t, p)$ are zero at the points of $\mathcal{R}$ over a_r; namely, $\lambda^{-1}(a_r) \setminus \{(r)\}$. Therefore the polar divisor of the meromorphic differential $(\zeta - a_r)^{-1}(\psi(\tau, t, p) \cdot u^r)\psi_\alpha(p)\Omega$. Suppose the Laurent series of ζ at (r) is given by

$$\zeta - a_r = \sum_{l=1}^{\infty} J_l^r z^{-l} \text{ at } (r) \tag{2.20a}$$

for some constants J_l^r. The formula

$$J_1^r = x_r \cdot u_r \tag{2.20b}$$

follows from the residue theorem and the definition of ρ_r and the formula,

$$\Omega = (\psi \cdot \phi)^{-1} d\zeta, \tag{2.20c}$$

holds to order 0 in z at each (r). We shall assume that (2.20) holds identically on $\mathcal{R}$.

(2.21) THEOREM: (Flaschka [9b].) Let $\Delta = \delta + (z)_0 - \infty$. The Baker function Ψ of Δ is given by this formula:

$$\Psi^r(t,p) = (\psi(0,t,p) \cdot u^r(0,t)) \frac{z^{-1}}{\zeta - a_r} \exp(-\theta(t,p)) \tag{2.21a}$$

or, in vector notation,

$$\Psi(t,p) = (\zeta - \mathbf{a})^{-1} \sum_{\gamma=1}^{\ell} \psi_\gamma(0,t,p) u_\gamma(0,t) z^{-1} \exp(-\theta(t,p)). \tag{2.21b}$$

The dual function is given by this formula:

$$\Phi(t,p) = (\zeta - \mathbf{a})^{-1} \sum_{\gamma=1}^{\ell} \phi_\gamma(0,t,p) x_\gamma(0,t) z^{-1} \exp(\theta(t,p)). \tag{2.21c}$$

Proof: The technical assumption preceeding our theorem implies that the function $\psi \cdot x^r$ has a zero at each point of $\mathcal{R}$ over a_r of order equal to that of $\zeta - a_r$. With this in mind, it is clear that the right hand side in (2.21a) is finite at the points of $\mathcal{R}$ over a_r. In a neighborhood of ∞ the right hand side is holomorphic (the exponential in ψ cancels with $\exp(-\theta)$) and

$$\Psi = u_\alpha(t)\zeta^{m-1} + O(\zeta^{m-2}) \text{ at } \infty_\alpha. \tag{2.21d}$$

The right hand side satisfies both (2.7a) and (2.7b). □

(2.21e) REMARK: If $m = 0$ then Ψ and Φ have simple poles at the ∞_α.

(2.21f) REMARK: We have made an harmless deviation from (2.7c) in that the value of $\Psi^r \exp(-\theta)$ at (r) is not 1 but ρ_r^{-1}.

Isospectral Matrices. We are now in a position to construct two $\mathbf{m}$-dependent matrices whose spectrum is preserved along the curve $t_j \to \mathbf{m}(t)$. The AKNS-Neumann parameter τ is set to 0 for this.

(2.22) THEOREM: The Baker function ψ satisfies these $\ell \times \ell$ matrix spectral problem:

$$\begin{cases} z(p)\psi(t,p) = Z(\mathbf{m}(t), \zeta(p))\psi(t,p) & \textbf{(2.22a)} \\ z(p)\zeta(p)^m \psi(t,p) = \tilde{Z}(\mathbf{m}(t), \zeta(p))\psi(t,p) & \textbf{(2.22b)} \end{cases}$$

where

$$\begin{cases} Z = \epsilon\delta_{m,0} + \sum_{r=1}^{\ell'} \frac{x^r \otimes u^r}{\zeta - a_r} = \epsilon\delta_{m,0} + \mathbf{x}^T(\zeta - \mathbf{a})^{-1}\mathbf{u} & \textbf{(2.22c)} \\ \tilde{Z} = \epsilon + \sum_{r=1}^{\ell'} a_r^m \frac{x^r \otimes u^r}{\zeta - a_r} = \epsilon + \mathbf{x}^T\mathbf{a}^m(\zeta - \mathbf{a})^{-1}\mathbf{u} & \textbf{(2.22d)} \end{cases}$$

The dual function satisfies the adjoint spectral problem and Z satisfies this Lax equation:

(2.22e)
$$\partial_{k,\alpha} Z = [B_{k,\alpha}, Z].$$

Proof: The function $(\zeta^m - \epsilon_\alpha z^{-1})\psi_\alpha \exp(-\theta)$ belongs to the linear space defined in (2.7). Its components in the basis Ψ^r are computed using (2.7c) with ρ_r in place of $\delta_{r,s}$. The component r in the following formula is verified by letting $p \to (r)$,

(2.22f)
$$(\zeta^m - \epsilon_\alpha z^{-1})\psi_\alpha \exp(-\theta) = \mathbf{a}^m x_\alpha \cdot \Psi.$$

These formulas are equivalent to the matrix equation (2.22b) using (2.21b). The equation (2.22a) has a similar proof. The Lax eqation (2.22e) is the integrability condition of the linear equations (2.22a) and (2.14). The dual equations may be derived from Cherednik's work or by applying the above analysis to ϕ. □

(2.23) PROPOSITION: There exists a unique $\ell \times \ell$ matrix $\mathbb{B}_{k,\alpha}$ of the form $\mathbb{B}_{k,\alpha} = \mathbf{a}^k z^\alpha +$ (terms of lower order in z) such that the vector Baker function and its dual satisfy these linear equations:

$$\begin{cases} \partial_{k,\alpha} \Phi(t,p) = \mathbb{B}_{k,\alpha}(\mathbf{m}(t), z(p))\Phi(t,p) \\ \text{and} \\ \partial_{k,\alpha} \Psi(t,p) = -\mathbb{B}_{k,\alpha}(\mathbf{m}(t), z(p))^T \Psi(t,p). \end{cases}$$

Proof: Let Φ_j is the $\ell' \times \ell'$ matrix whose column s is defined by the Taylor series of $\Phi \exp(-\theta)$ at (s):

$$\Phi = \rho_s^{-1}(\sum_{j=0}^{\infty} \Phi_j^{(s)} z^{-j}) \exp(\theta).$$

As in the construction of ψ_∞, let

$$\Phi_\infty = \sum_{j=0}^{\infty} \Phi_j z^{-j} \exp(\Theta)\rho^{-1}$$

where

$$\Theta = \sum_{W} \Lambda^k z^\alpha t_{k,\alpha}, \quad \Lambda = \mathbf{a} + \sum_{\beta=1}^{\infty} J_\beta z^{-\beta}, \quad \rho = \operatorname{diag}(\rho_r) \text{ and } J_\beta = \operatorname{diag}(J_\beta^r).$$

A series of matrices $\mathbb{B}^{(\eta)}$ is defined in an iterative way by the these formula:

$$\mathbb{B}^{(0)} = \mathbf{a}^k \quad \text{and} \quad \mathbb{B}^{(\eta)} = T_k^{(\eta)} + \sum_{\beta=0}^{\eta-1} (\Phi_{\eta-\beta} T_k^{(\beta)} - \mathbb{B}^{(\beta)} \Phi_{\eta-\beta})$$

where $\Lambda^k = \sum_{\beta \geq 0} T_k^{(\beta)} z^{-\beta}$. The matrix $\mathbb{B} = \sum_{\eta=0}^{\alpha} \mathbb{B}^{(\eta)} z^{\alpha-\eta}$ satisfies

$$(\partial_{k,\alpha} \Phi_\infty - \mathbb{B}\Phi_\infty) \exp(-\Theta) = \sum_{j \geq 0} \Phi_j z^{-j} \Lambda^k z^\alpha - \mathbb{B} \sum_{j \geq 0} \Phi_j z^{-j} + O(z^{-1})$$

$= O(z^{-1})$. This implies that $(\partial_{k,\alpha} \Phi - \mathbb{B}\Phi_) \exp(-\theta) = O(z^{-1})$ at each (s). Since the components of Φ are a basis for the linear space defined in (2.7), the matrix $\mathbb{B}_{k,\alpha} = \mathbb{B}$ satisfies the first formula of our proposition. It is clear from our construction that $\mathbb{B}$ is unique. According to Cherednik [6], the dual function satisfies the adjoint equations. □

(2.24) REMARK: The $\mathbf{m}$-dependence of our matrices $\mathbb{B}_{k,\alpha}$ is at this point implicit. This situation is remedied by the following construction.

(2.25) THEOREM: The Baker function satisfies an $\ell' \times \ell'$ matrix spectral problem of the form:

$$\lambda(p)\zeta(p)\Phi(t,p) = \mathbb{K}(\mathbf{m}(t), z(p))\Phi(t,p)$$

where $\lambda = \prod_{\alpha=1}^{\ell}(z - \epsilon_\alpha)$ if $m = 0$, $\lambda = z$ if $m > 0$ and $\mathbb{K}$ is a matrix polynomial in z of degree ℓ if $m = 0$ or 1 if $m > 0$.

Proof: The result follows, according to [12], from the fact that λ is holomorphic in $\mathcal{R} \setminus \infty$. □

(2.26) THEOREM: The Baker function Φ satisfies these $\ell' \times \ell'$ matrix spectral problem:

$$\begin{cases} \zeta(p)\Phi(t,p) = \mathbb{L}(\mathbf{m}(t), z(p))\Phi(t,p) & \textbf{(2.26a)} \\ \zeta(p)\Phi(t,p) = \tilde{\mathbb{L}}(\mathbf{m}(t), \zeta(p)^m z(p))\Phi(t,p) & \textbf{(2.26b)} \end{cases}$$

where

$$\begin{cases} \mathbb{L} = \mathbf{a} + \sum_{\alpha=1}^{\ell} \frac{x_\alpha \otimes u_\alpha}{z - \epsilon_\alpha \delta_{m,0}} = \mathbf{a} + \mathbf{x}(z - \epsilon\delta_{m,0})^{-1}\mathbf{u}^T & \textbf{(2.26c)} \\ \tilde{\mathbb{L}} = \mathbf{a} + \sum_{\alpha=1}^{\ell} \frac{x_\alpha \otimes u_\alpha}{\zeta^m z - \epsilon_\alpha} \mathbf{a}^m = \mathbf{a} + \mathbf{x}(\zeta^m z - \epsilon)^{-1}\mathbf{u}^T\mathbf{a}^m & \textbf{(2.26d)} \end{cases}$$

The dual function satisfies the adjoint spectral problem and $\mathbb{L}$ satisfies this Lax equation:

$$\textbf{(2.26e)} \qquad \partial_{k,\alpha}\,\mathbb{L} = [\mathbb{B}_{k,\alpha}, \mathbb{L}].$$

Proof: The formulas may be verified directly by applying $\mathbb{L}$ to Φ using (2.22f) and (2.21b). □

(2.26f) REMARK: The matrix $\mathbb{K}$ in (2.25) is given by this formula: $\mathbb{K} = \lambda\mathbb{L}$.

(2.27) THEOREM: The matrix $\mathbb{B}_{k,\alpha}$ is given explicitely by the formula $\mathbb{B}_{k,\alpha} = \pi_+(z^\alpha \mathbb{L}^k)$ where π_+ denotes projection onto the polynomial part of its argument.

Proof: The formula (2.26a), in the notation used in the proof of proposition (2.23), is equivalent to this one:

$$\Phi_\infty \Lambda = (\mathbb{L})\Phi_\infty \quad \text{and} \quad \Phi_\infty \Lambda^k = \mathbb{L}^k \Phi_\infty.$$

Now it is clear that $\partial_{k,\alpha}\Phi - \pi_+(z^\alpha \mathbb{L}^k)\Phi = O(z^{-1})$ at each ∞_α and this implies the first formula in proposition (2.23). □

Proposition (2.23) with $p \to \infty_\gamma$ leads to another representation of the equations of motion (2.18). Using (2.21d) and its Φ-analogue, one finds that

$$\textbf{(2.28)} \qquad \partial_{k,\alpha}\,\mathbf{m} = X_{k,\alpha}(\mathbf{m}) : \begin{cases} \partial_{k,\alpha}\, x_\gamma = \mathbb{B}_{k,\alpha}(\mathbf{m}(t), z(\infty_\gamma))x_\gamma \\ \partial_{k,\alpha}\, u_\gamma = -\mathbb{B}(\mathbf{m}(t), z(\infty_\gamma))^T u_\gamma, \end{cases}$$

where $z(\infty_\gamma) = \epsilon_\gamma$ if $m = 0$ and $z(\infty_\gamma) = 0$ if $m > 0$.

(2.29) REMARK: The last development, which culminated with the formula (2.28), was based on the relationship between $\mathbb{L}$ and z. There is an alternative development based on the variables $\tilde{\mathbb{L}}$ and $\tilde{z}$ where $\tilde{z} \doteq \zeta^m z$. The results are summarized by the following analogues of the above formulas:

$$\theta(t,p) = \sum_W \zeta^k z^\alpha t_{k,\alpha} = \sum_W \zeta^{k-m\alpha} \tilde{z}^\alpha t_{k,\alpha}$$

$$\left\{\begin{array}{l}\partial_{k,\alpha}\,\Phi(t,p) = \tilde{\mathbb{B}}_{k,\alpha}(\mathbf{m}(t),\tilde{z}(p))\Phi(t,p)\\ \text{and}\\ \partial_{k,\alpha}\,\Psi(t,p) = -\tilde{\mathbb{B}}_{k,\alpha}(\mathbf{m}(t),\tilde{z}(p))^T\Psi(t,p).\end{array}\right. \tag{2.23'}$$

$$\tilde{\mathbb{B}}_{k,\alpha} = \pi_+(\tilde{z}^\alpha \tilde{\mathbb{L}}^{k-m\alpha}) \tag{2.27'}$$

$$\partial_{k,\alpha}\,\mathbf{m} = X_{k,\alpha}(\mathbf{m}) : \quad \left\{\begin{array}{l}\partial_{k,\alpha}\,x_\gamma = \tilde{\mathbb{B}}_{k,\alpha}(\mathbf{m}(t),\epsilon_\gamma)x_\gamma\\ \\ \partial_{k,\alpha}\,u_\gamma = -\tilde{\mathbb{B}}_{k,\alpha}(\mathbf{m}(t),\epsilon_\gamma)^T u_\gamma.\end{array}\right. \tag{2.28'}$$

CHAPTER III

THE DIVISOR MAP

The Isospectral Curve. Let M be the submanifold of $\mathbb{C}^{2\ell\ell'}$ defined in (2.15) and let $\mathbf{m} \in$ M. Let $\mathbf{m}(\tau)$ be the integral curve of the vector field X in (2.17) through $\mathbf{m}$. Our purpose in this section is to reproduce $\mathbf{m}(\tau)$ by the construction used in the previous section. We shall need to make two assumptions concerning $\mathbf{m}$. It is assumed that both x^r and u^r have rank 1 for each $(r = 1, \dots, \ell')$. There is no loss of generality in this assumption for if some x^r is 0 then $x^r(\tau)$ is identically zero. The set of all such points, denoted M^1 as in Chapter I, is a dense open submanifold of M. The second assumption is explained in the following paragraphs.

Let $\mathcal{R}$ be the algebraic curve defined by the characteristic equation of the matrix Z in theorem (2.22):

$$\mathcal{R}: \quad g(\zeta, z) = |Z(\mathbf{m}, \zeta) - zI_n| = 0$$

where

$$Z = \epsilon\delta_{m,0} + \sum_{r=1}^{\ell'} \frac{x^r \otimes u^r}{\zeta - a_r} = \epsilon\delta_{m,0} + \mathbf{x}^T(\zeta - \mathbf{a})^{-1}\mathbf{u} \tag{3.1a}$$

More precisely, $\mathcal{R}$ is the projective curve with affine part $\mathcal{R}_0 = \{(\zeta, z) \in \mathbb{C}^2 : g(\zeta, z) = 0\}$ plus points where one of ζ or z is infinite. We shall refer to $\mathcal{R}$ as the *spectrum* of $\mathbf{m}$. The set $M(\mathcal{R})$ of all points $\mathbf{m}$ in M whose spectrum is $\mathcal{R}$ is called the *isospectral surface* of $\mathbf{m}$. Our purpose now is to show that z, thought of as a meromorphic function on $\mathcal{R}$, satisfies the three conditions listed in (2.1).

Let us examine the spectral problem (2.22a) with $t = 0$ in a complex neighborhood of $\zeta = a_r$. The Laurent series of Z in ζ at a_r has this form:

$$Z = \frac{x^r \otimes u^r}{\zeta - a_r} + \sum_{j=0}^{\infty} Z_j^r(\mathbf{m})(\zeta - a_r)^j$$

for some matrices Z_j^r. The only possibly nonzero eigenvalue of $x^r \otimes u^r$ is $x^r \cdot u^r$ and the corresponding eigenvector is x^r. The eigenspace corresponding to the zero eigenvalue consists of all vectors perpendicular to u^r. A straightforward calculation shows that the spectral problem has a solution in the following form:

$$\textbf{(3.1b)} \qquad U = x^r + \sum_{j=1}^{\infty} U_j^r z^{-j} \quad \text{with} \quad \zeta - a_r \doteq \sum_{k=1}^{\infty} J_k^r z^{-k}.$$

One may show, by substituting U for ψ in (2.22a), that

$$J_1^r = x^r \cdot u^r, \quad U_1^r = \epsilon\delta_{m,0} x^r + \sum_{s \neq r} \frac{x^r \cdot u^s}{(a_r - a_s)} x^s, J_2^r = x^r \cdot U_1^r$$

and so on. This proves that $\mathcal{R}$ contains a point (r) over a_r where z has a simple pole. As before, z^{-1} shall serve as a local parameter at (r). The characteristic equation (3.1a) of Z must have the following form:

$$\textbf{(3.1c)} \qquad g(\zeta, z) = (-z)^\ell + \sum_{\alpha=1}^{\ell} \left(\sum_{r=1}^{\ell'} \frac{1}{\zeta - a_r} G_\alpha^r \right) z^{\ell - \alpha} = 0$$

for some functions G_α^r of $\mathbf{m}$. Let g_α denote the coefficient of $z^{\ell-\alpha}$.

Next we examine $\mathcal{R}$ over $\zeta = \infty$. The Taylor series of Z at $\zeta = \infty$ is given by this form:

$$Z = \epsilon\delta_{m,0} + \sum_{j=1}^{\infty} \mathbf{x}^T \mathbf{a}^{j-1} \mathbf{u} \zeta^{-j}$$

In view of the constraints (2.15a), Z is given by this formula:

$$\textbf{(3.1d)} \qquad Z = \zeta^{-m}(\epsilon + \mathbf{x}^T \mathbf{a}^m (\zeta - \mathbf{a})^{-1}) \mathbf{u}) = \zeta^{-m} \tilde{Z}$$

where $\tilde{Z}$ was defined in (2.22d). Now it is easy to see that the equation $g(\zeta, z) = 0$ has a solution of the form

$$\textbf{(3.1e)} \qquad z = \zeta^{-m}(\epsilon_\alpha + \sum_{j=1}^{\infty} z_{\alpha,j} \zeta^{-j}).$$

This means that $\mathcal{R}$ contains just ℓ points at $\zeta = \infty$. We call these points $\infty = \infty_+ \cdots + \infty_\ell$.

Functional Relations. We have at this point introduced two sets of functions, the G^r_α and the J^r_k. It will turn out that these functions are constants of motion; that is, they are constant along all integral curves of the vector fields (2.17) and (2.29). It is clear that these functions are constant along the particular integral curve constructed in Chapter II. For now we wish to point out that these functions are not independent. They are related by the following formula:

$$0 = g(\zeta, z) = g(a_r + \sum_{j=1}^{\infty} J^r_\alpha z^{-j} \, , \, z). \tag{3.1f}$$

It convenient to introduce at this point a third set of functions, the $H_{k,\alpha}$. Let

$$h(\zeta, z) \doteq h_0(\zeta) g(\zeta, z) = h_0(\zeta)(-z)^\ell + \sum_{\alpha=1}^{n} h_\alpha(\zeta) z^{\ell-\alpha}$$

where

$$h_0 \doteq \prod_{r=1}^{\ell} (\zeta - a_r) \quad \text{and} \quad h_\alpha = h_0 g_\alpha.$$

The Weinstein-Aronszan formula [14]applied to the formulas (2.22c) and (2.26c) is this:

$$\begin{aligned} |\mathbb{L} - \zeta I| &= |\mathbf{a} - \zeta I| \cdot |I + (\mathbf{a} - \zeta)^{-1} \mathbf{x} (zI - \epsilon \delta_{m,o})^{-1} \mathbf{u}^T| \\ &= |\mathbf{a} - \zeta I| \cdot |I + (zI - \epsilon \delta_{m,o})^{-1} \mathbf{x}^T (\mathbf{a} - \zeta)^{-1} \mathbf{u}| \\ &= |\mathbf{a} - \zeta I| \cdot |zI - \epsilon \delta_{m,o}|^{-1} \cdot (-1)^\ell |Z - zI| \\ &= (-1)^{\ell + \ell'} |zI - \epsilon \delta_{m,o}|^{-1} h(\zeta, z). \end{aligned} \tag{3.1g}$$

The matrices $\tilde{\mathbb{L}}$ and $\tilde{Z}$ in (2.22d) and (2.26d) are related in a similar way:

$$|\tilde{\mathbb{L}} - \zeta I| = (-1)^\ell |\mathbf{a} - \zeta I| \cdot |z \zeta^m I - \epsilon|^{-1} \cdot |\tilde{Z} - z \zeta^m I|. \tag{3.1h}$$

The h_α are polynomials in ζ with $\mathbf{m}$ dependent coefficients, the $H_{k,\alpha}$. The formula (3.1g) contains the relationship between these coefficients and the G^r_α.

There are linear relationships among the G^r_α themselves; namely,

$$\sum_{r}^{\ell} a^j_r G^r_\alpha = 0 \text{ if } (j = 0, \ldots, m\alpha - 1). \tag{3.1i}$$

These are obtained by substituting (3.1e) into (3.1c) with ζ tending to ∞. The number of these relations, $\frac{1}{2} m\ell(\ell - 1)$, is one-half the number of constraints (2.15a).

(3.1j) REMARK: It is clear that the isospectral surface $M(\mathcal{R})$ is simply the level set of the G^r_α through **m**. The number of functions among the G^r_α that are independent on $M(\mathcal{R})$ is at most:

$$\ell\ell' - \frac{1}{2}m\ell(\ell-1) = g_{\mathcal{R}} + \ell + \ell' - 1$$

Therefore the dimension of $M(\mathcal{R})$ is, by (2.15b), at least

$$2(g_{\mathcal{R}} + \ell + \ell' - 1) - (g_{\mathcal{R}} + \ell + \ell' - 1) = g_{\mathcal{R}} + \ell + \ell' - 1.$$

We assumpe that $\mathcal{R}$ is a smooth algebraic curve. This assumption is going to be used below to prove that all functional relationships among the G^r_α are included in the count (7.1i) and $M(\mathcal{R})$ is a level set of dimension $g_{\mathcal{R}} + \ell$. The vanishing of $\nabla_{(\zeta,\zeta)}g$, which happens in the presence of singularities, would amount to more functional relationships among the G^r_α.

Genus. It is an elementary fact that the zero divisor of the abelian differential $d\zeta$, restricted to $\mathcal{R}_0$, is equal to the zero divisor of the discriminant $\partial h/\partial z$ restricted to $\mathcal{R}_0$:

$$(d\zeta)_0|_{\mathcal{R}_0} = (\partial h/\partial z)_0|_{\mathcal{R}_0}.$$

The poles of $\partial h/\partial z$ lie among the (r) and ∞ and their number is equal to the number of zeros of $\partial h/\partial z$; hence,

$$\deg((\partial h/\partial z)_0|_{\mathcal{R}_0}) = -\sum_{\alpha=1}^{\ell} \mathrm{ord}_{\infty_\alpha}(\partial h/\partial z) - \sum_{r=1}^{\ell'} \mathrm{ord}_{(r)}(\partial h/\partial z)$$

and so

$$\deg((\partial h/\partial z)_0|_{\mathcal{R}_0}) = \deg(\partial h/\partial z)_\infty.$$

Now 2∞ is the polar divisor of $d\zeta$. Using this and the last two formulas one may conclude that

$$\deg((d\zeta)_0) = \deg((d\zeta)_0|_{\mathcal{R}_0}) + \sum_{r=1}^{\ell'} \mathrm{ord}_{(r)}(d\zeta)$$

$$= - - \sum_{\alpha=1}^{\ell} \mathrm{ord}_{\infty_\alpha}(\partial h/\partial z) + \sum_{r=1}^{\ell'} \mathrm{ord}_{(r)}((\partial h/\partial z)^{-1}d\zeta).$$

The differential $(\partial h/\partial z)^{-1}d\zeta$ has a zero at (r) of order $\ell-2$ and a pole at ∞_α of order equal to

$$-\operatorname{ord}_\infty(h_0(\zeta)z^{\ell-1}) = \ell' - m(\ell-1).$$

The last two formulas imply that

$$\deg((d\zeta)_0) = \ell(\ell' - m(\ell-1)) + (\ell-2)\ell'.$$

The Riemann-Hurwitz formula, $\deg(d\zeta) = 2g_{\mathcal{R}} - 2$, implies that

$$g_{\mathcal{R}} = (\ell-1)(\ell'-1) - \frac{1}{2}m\ell(\ell-1). \tag{3.2}$$

This proves that (2.1c) is the correct formula for the genus of $\mathcal{R}$.

Holomorphic Differentials. It is an elementary fact that the linear space $H^1(\mathcal{R})$ of holomorphic differentials on $\mathcal{R}$ has dimension $g_{\mathcal{R}}$ over $\mathbb{C}$ and it is generated over $\mathbb{C}[\mathcal{R}]$ by $\eta \doteq (-1)^\ell \ell(\partial h/\partial z)^{-1}d\zeta$. The generator has a zero at ∞_α of order $\operatorname{ord}_{\infty_\alpha}(h_0(\zeta)z^{\ell-1}) - 2 = \ell' - m(\ell-2) - 2$ and a zero at each (r) of order $\ell-2$:

$$\eta = \begin{cases} \frac{1}{\epsilon_\alpha^{\ell-1}\zeta^{\ell'-m(\ell-1)-2}}(-1+O(\zeta^{-1}))d\zeta^{-1} & \text{at } \infty_\alpha \\ \frac{b_r}{z^{\ell-2}}\prod\limits_{s\neq r}(a_r-a_s)^{-1}(1+O(z^{-1}))dz^{-1} & \text{at (r).} \end{cases} \tag{3.3}$$

where $b_r = \operatorname{ord}_r(\zeta - a_r)$. It follows then that $H^1(\mathcal{R})$ is generated over $\mathbb{C}$ by the differentials $z^\alpha\zeta^r\eta$ where $(\alpha = 0,\dots,\ell-2)$ and $(r = 0,\dots,\ell' - m(\ell-1-\alpha)-2)$.

The Divisor ∞. The set of Weierstrass gap numbers of ∞ (See Chapter II.) has a surprisingly simple structure. Let $\mathfrak{A}_\infty$ be the ring of meromorphic functions holomorphic in $\mathcal{R}\setminus\infty$. Define for each $(j = 1,\dots,\ell-1)$

$$\lambda_j \doteq (-1)^\ell h_0(\zeta)z^{\ell-j} + \sum_{k=1}^{\ell-j} h_k(\zeta)z^{\ell-j-k}.$$

The next formula follows from $h(\zeta,z) = 0$:

$$\lambda_j = -\sum_{l=\ell-j+1}^{\ell} h_l(\zeta)z^{\ell-j-l}.$$

The first formula suggests that the poles of λ_j lie in $(z)_\infty + \infty$ while the second formula suggests that they lie in $(z)_0 + \infty$. Thus any pole of λ_j lies in ∞ and therefore λ_j belongs to $\mathfrak{A}_\infty$.

The Divisor of m. Let $\mathbf{m} \in M(\mathcal{R})$. The divisor of **m**, denoted δ_m is defined in terms of the spectral problem (2.22a) with $t = 0$. It has, for most points p of $\mathcal{R}$, a solution U(p), the sum of whose components is 1. The components of U are rational functions on $\mathcal{R}$. Let δ_m be the minimal positive divisor satisfying this condition:

(3.4) $$\delta_m + (U_\alpha) \geq 0 \text{ for each } (\alpha = 1, \ldots, \ell).$$

Our immediate purpose is to show that δ_m satisfies the condition (2.2a) and that its degree is $g_{\mathcal{R}} + \ell - 1$.

The spectral problem (2.22a) at $p = \infty_\alpha$ is given by $\epsilon U(\infty_\alpha) = \epsilon_\alpha U(\infty_\alpha)$. The solution U is given in a neighborhood of ∞_α by a series of this form:

(3.5) $$U = (e_\alpha + \xi_1^\alpha \zeta^{-1} + \cdots).$$

This series may be substituted into (2.22a). The computation leads to iterative formulas for the ξ_j^α. The following lemma is a consequence of (3.5).

(3.6) LEMMA: (3.6a) The degree of δ_m is $g_{\mathcal{R}} + \ell - 1$. **(3.6b)** $L(\delta_m - \infty) = \{0\}$.

Proof: Let $\Delta_{\alpha,\beta}$ denote the cofactor of entry (α, β) of the matrix $Z - zI$. These cofactors satisfy this formula:

(3.6c) $$\frac{\Delta_{\alpha,\beta}}{\Delta_{\alpha,\gamma}} = \frac{\Delta_{\epsilon,\beta}}{\Delta_{\epsilon,\gamma}},$$

and the component β of U is given for any α by this formula:

(3.6d) $$U_\beta = \frac{\Delta_{\alpha,\beta}}{\sum_{\gamma=1}^{\ell} \Delta_{\alpha,\gamma}}.$$

If a component of U has a pole at a point p of $\mathcal{R}$ then

(3.6e) $$\sum_{\gamma=1}^{\ell} \Delta_{\alpha,\gamma} = 0$$

for each $(\alpha = 1, \ldots, \ell)$. Now each $\Delta_{\alpha,\alpha}$ is a polynomial in z of degree $\ell - 1$ and each $\Delta_{\alpha,\beta}$, $(\alpha \geq \beta)$ is a polynomial in z of degree $\ell - 2$. The *polar conditions* (3.7c), may be thought of as an $(\ell - 1) \times (\ell - 1)$ system of linear equations for the vector

$(1, z, \cdots, z^{\ell-1})^T$. The determinant of the coefficient matrix is a rational function $\mathcal{P}$ of ζ. The ζ-projection of any solution p to (3.7c) is a root of $\mathcal{P}$. Thus the degree of δ_m is no greater then the number of roots of $\mathcal{P}$.

The order of the pole of $\mathcal{P}(\zeta)$ at a_r may be computed by examining the polar conditions (3.7c) with ζ approaching a_r. The $z^{\ell-\alpha}$ term of a cofactor is a product of $\alpha - 1$ entries of Z but the order of its pole at a_r is at most 1; otherwise, some g_α would have a double pole or worse. Thus the order of the pole of $\mathcal{P}$ at a_r is at most $\ell - 1$.

Let us compute the order of the zero of $\mathcal{P}(\zeta)$ at ∞. In view of (3.1d), the polar conditions are given at $\zeta = \infty$ by this formula:

$$\sum_{\gamma=1}^{\ell} \Delta_{\alpha,\gamma} = \Delta_{\alpha,\alpha} + \cdots = \zeta^{-m(\ell-1)} \prod_{\gamma\neq\alpha} (\epsilon_\beta - z)$$

plus terms of lower order in ζ. The order of the zero at ∞ of $\mathcal{P}$ can now be computed directly. It is one-half $\times$ the number of constraints: $\frac{1}{2}\ell(\ell-1)m$.

The number of zeros of $\mathcal{P}$ away from $\zeta = \infty$ is $(\ell-1)\ell' - \ell(\ell-1)m/2 = g_{\mathcal{R}} + \ell - 1$. (If $\mathcal{P}$ has a pole at a_r of order less than ℓ then a_r is counted as a zero.) Thus there exists a divisor $\mu = \mu_1 + \cdots + \mu_{g_{\mathcal{R}}+\ell-1}$ over $\mathbb{P}^1$ such that

$$\text{(3.6f)} \qquad \mathcal{P}(\zeta) = \frac{\prod_{j=1}^{g_{\mathcal{R}}+\ell-1} (\zeta - \mu_j)}{\prod_{r=1}^{\ell'} (\zeta - a_r)^{\ell-1}}.$$

It has yet to be shown that the degree of δ_m is $g_{\mathcal{R}} + \ell - 1$. The arguement, which is taken from [18b], is this. If each $\sum_{\gamma=1}^{\ell} \Delta_{\alpha,\gamma}, (\alpha = 1, \ldots, n)$ vanishes to order s in some local parameter at p and if each component of U has a pole of order less than s at p then by (3.6d) each $\Delta_{\alpha,\beta}$ must vanish at p. This would imply that the functions

$$\text{(3.6g)} \qquad \frac{\partial g}{\partial z} = -\sum_{\alpha=1}^{n} \Delta_{\alpha,\alpha} \quad \text{and} \quad \frac{\partial g}{\partial \lambda} = \sum_{\alpha,\beta} \Delta_{\alpha,\beta} \frac{\partial Z_{\alpha,\beta}}{\partial \lambda}$$

vanish at p; p is a singular point. Since $\mathcal{R}$ is a smooth curve, the multiplicity of p in δ_m is at least s. Now suppose that the degree of δ_m is strictly less than $g_{\mathcal{R}} + \ell - 1$. Then some μ comes from a solution to the pole conditions that does not contribute a pole to δ_m. According to the previous paragraph, this cannot happen. The degree of δ_m is $g_{\mathcal{R}} + \ell - 1$.

To prove the second assertion, let s be an integer such that $\delta_m + s\infty$ is a nonspecial divisor. Then, by the Riemann-Roch theorem and the previous assertion, the dimension of $L(\delta_m + s\infty)$ is $s(\ell + 1)$. The set of functions of the form $\zeta^j U_\alpha, j \geq 0, (\alpha = 1, \dots, \ell)$ is linearly independent. Thus $L(\delta_m + s\infty)$ has a basis coming from these functions. Therefore the U_α are a basis for $L(\delta_m)$ and (7.6b) follows. □

(3.7) REMARK: It is known that the Hamilton-Jacobi equation of the original Neumann system separates in the μ-coordinate system on S^ℓ; see [14].

Dual Divisor. The dual divisor δ^ι_m, which is defined by Propositon (2.4), has a concrete description in terms of the transpose of the spectral problem (2.22a). The transpose problem has, for all but a finite number of points of $\mathcal{R}$, a solution V(p) whose components sum to 1. The components of V are rational functions given for any β by this formula:

$$V = \frac{(\Delta_{1,\beta}, \dots, \Delta_{n,\beta})^T}{\sum_{\gamma=1}^{\ell} \Delta_{\gamma,\beta}}.$$

It is given in a neighborhood of $\infty - \alpha$ by this formula:

(3.8a)
$$V = e_\alpha + O(\zeta^{-1}.$$

The function $U \cdot V$ has poles in $\delta + (V)_\infty$. The next formula follows from (3.6c) and (3.6g):

(3.8b)
$$U \cdot V = -\frac{1}{\sum_{\alpha,\beta=1}^{\ell} \Delta_{\alpha,\beta}} \frac{\partial g}{\partial z}.$$

(3.8c)
$$(U \cdot V)^{-1} d\zeta = \frac{(-1)^{\ell+1}}{\ell} h_0(\zeta) \sum_{\alpha,\beta=1}^{\ell} \Delta_{\alpha,\beta}\eta$$

With this, the next two formulas follow by Proposition (2.4):

(3.8d)
$$\Omega = (U \cdot V)^{-1} d\lambda \quad \text{and} \quad \delta^\iota = (V)_\infty.$$

The Solution. Let $\mathbf{m}$ be a point or $M(\mathcal{R})$. The construction of Chapter II may be used in the following way to find the integral curve of the AKNS/Neumann system through $\mathbf{m}$. Let $\delta = \delta_m$ be the divisor of $\mathbf{m}$ and let δ^ι be its dual divisor. Let ψ be the Baker function of δ and let ϕ be the dual function. Using (3.5), (2.6c), the Riemann-Roch theorem and the fact that both of U_α and $\psi_\alpha(0,0,p)$ belong to $L(\delta)$, it is easy to argue that $\psi(0,0,p) = U(p)$. Let $\mathbf{m}(\tau,t)$ be the point of $\mathbb{C}^{2n\ell}$ defined by the formulas preceeding (2.15). Then $\mathbf{m}(\tau,t)$ satisfies the equations (2.17) and (2.29) of the generalized AKNS/Neumann hierarchy. The formula $\mathbf{m}(0,0) = \mathbf{m}$ follows now from (7.1b). The solution method has several important implications.

(3.9) THEOREM: The generalized AKNS/Neumann vector fields X^* and $X^*_{k,\alpha}$, (2.17), (2.14) and (2.29), are tangential to M and $M(\mathcal{R})$. The $g_{\mathcal{R}}$ vector fields corresponding to the Weierstrass gap numbers of ∞ are linearly independent at each point of $M(\mathcal{R})$.

Proof: The vector fields are tangential to M by proposition (2.19). The construction of Chapter II leads to a pair of matrices Z and $\mathbb{L}$ whose spectra are constant along the integral curve $\mathbf{m}(\tau,t)$. Now it is clear that their spectra are preserved along **all** integral curves on M. The vector fields X^* and $X^*_{k,\alpha}$ are tangential to $M(\mathcal{R})$ because $M(\mathcal{R})$ is defined by the characteristic polynomial of Z.

Suppose there is a linear dependence relation among the $X^*_{k,\alpha}(\mathbf{m})$; say,

$$\sum_W c_{k,\alpha} X^*_{k,\alpha}(\mathbf{m}) = 0.$$

Let B be the differential operator in τ defined by this formula:

$$B \doteq \sum_W c_{k,\alpha} L_{k,\alpha}$$

where the $L_{k,\alpha}$ are defined by (2.10a). Let $y = (B\psi)_1 \cdot \psi_1^{-1}$. One one hand, for $\tau \neq 0$, y is a rational function whose poles lie in ∞; that is, $y \in \mathfrak{A}_\infty$. On the other hand, in view of Krichever's homomorphism [12], if $y \neq 0$ then $\deg((y)_\infty)$ is a Weierstrass gap number of ∞. This can only mean that $y = 0$. If at least one of the constants $c_{k,\alpha}$ is nonzero then B is a differential operator of positive degree whose leading coefficient is constant in (τ,t). Therefore $y = (B\psi)_1 \cdot \psi_1^{-1}$ goes like a positive power of ζ at some ∞_α. The contradiction is clear. All the $c_{k,\alpha}$ must be zero and the $X^*_{k,\alpha}(\mathbf{m})$ are independent vectors. □

(3.10) DEFINITION: Let Y^r denote the vector field defined by these equations:

$$\dot{\mathbf{m}} = Y^r(\mathbf{m}): \quad \dot{x}^s = x^r \delta_{r,s} \quad \dot{u}^s = -u^r \delta_{r,s}, \quad (r = 1, \dots, \ell'),$$

and let W_α denote the vector field defined by these equations:

$$\dot{\mathbf{m}} = W_\alpha(\mathbf{m}): \quad \dot{x}_\beta = x_\alpha \delta_{\alpha,\beta} \quad \dot{u}_\beta = -u_\beta \delta_{\alpha,\beta}, \quad (\alpha = 1, \dots, \ell).$$

Suppose $\theta_1, \dots, \theta_{\ell'}$ are constants. Let $\theta = \operatorname{diag}(\theta_1, \dots, \theta_{\ell'})$. and let $Y = \sum_{r=1}^{\ell'} \theta_r Y^r$. Let $\mathbf{m}(t)$ be the integral curve of Y through $\mathbf{m}$. Under the flow induced by Y one has:

$$\text{(3.10a)} \qquad \mathbf{x}(t) = e^{t\theta}\mathbf{x}, \quad \mathbf{u}(t) = e^{-t\theta}\mathbf{u} \quad \text{and} \quad Z(\mathbf{m}(t), \zeta) = Z(\mathbf{m}, \zeta).$$

The last formula implies that $U(\mathbf{m}(t)) = U(\mathbf{m})$ and δ_m is the divisor of $\mathbf{m}(t)$. Conversely, suppose two points $\mathbf{m}$ and $\tilde{\mathbf{m}}$ correspond to one the same divisor. Then their normalized eigenvectors must be equal, $U = \tilde{U}$ and $V = \tilde{V}$; hence,

$$\frac{1}{\sum_{\alpha=1}^{\ell} x_\alpha^r} x^r = \frac{1}{\sum_{\alpha=1}^{\ell} \tilde{x}_\alpha^r} \tilde{x}^r \quad \text{and} \quad \frac{1}{\sum_{\alpha=1}^{\ell} u_\alpha^r} u^r = \frac{1}{\sum_{\alpha=1}^{\ell} \tilde{u}_\alpha^r} \tilde{u}^r.$$

This implies that

$$\tilde{x}^r = e^{c_r} x^r \quad \text{and} \quad \tilde{u}^r = e^{c_r'} u^r$$

for some constants c_r and c_r'. Now $\mathbf{m}$ and $\tilde{\mathbf{m}}$ belong to $M(\mathcal{R})$ so $G_1^r(\mathbf{m}) = x^r \cdot u^r = G_1^r(\tilde{\mathbf{m}})$ and therefore $c_r' = -c_r$. This proves that $\tilde{\mathbf{m}}$ and $\mathbf{m}$ are equivalent under a flow generated by the X^r, $\tilde{\mathbf{x}} = e^\theta \mathbf{x}$ and $\tilde{\mathbf{u}} = e^{-\theta}\mathbf{u}$.

Suppose $\phi_1, \dots, \phi_\ell$ are constants. Let $\phi = \operatorname{diag}(\phi_1, \dots, \phi_\ell)$ and let $W = \sum_{\alpha=1}^{\ell} \phi_\alpha W_\alpha$. Let $\mathbf{m}(s)$ be the integral curve of W through $\mathbf{m}$. Under the flow induced by W one has:

$$\text{(3.10b)} \qquad \mathbf{x}(s) = \mathbf{x}e^{s\phi}, \quad \mathbf{u}(s) = \mathbf{u}e^{-s\phi}, \quad \text{and} \quad Z(\mathbf{m}(s), \zeta) = e^{s\phi} Z(\mathbf{m}, \zeta) e^{-s\phi}.$$

The last formula implies that

$$\text{(3.10c)} \qquad U(\mathbf{m}(s)) = \frac{\sum_{\gamma=1}^{\ell} \Delta_{\alpha,\gamma}}{\sum_{\gamma=1}^{\ell} e^{s\phi_\gamma} \Delta_{\alpha,\gamma}} e^{s\phi} U(\mathbf{m})$$

and therefore the divisor of $\mathbf{m}(s)$ is linearly equivalent to δ_m; they correspond to the same point of $\mathcal{J}(\mathcal{R})$. Conversely, suppose $\mathbf{m}$ and $\tilde{\mathbf{m}}$ are points of $M(\mathcal{R})$ corresponding to the same point of $\mathcal{J}(\mathcal{R})$; that is, $\mathcal{A}(\delta_m) = \mathcal{A}(\delta_{\tilde{\mathbf{m}}})$. Then by Abel's theorem, δ_m and $\delta_{\tilde{\mathbf{m}}}$ are linearly equivalent. Suppose f is a meromorphic function such that $(f) = \delta_m - \delta_{\tilde{\mathbf{m}}}$. Let U be the solution to (2.22a) corresponding to $\mathbf{m}$ and satisfying (3.5). Then

$$\textbf{(3.10d)} \qquad f = \sum_{\alpha=1}^{\ell} f(\infty_\alpha) U_\alpha \quad \text{and} \quad \tilde{U}(p) = \frac{\operatorname{diag}(f(\infty_\alpha))}{f(p)} U(p)$$

where the constants $f(\infty_\alpha)$ are nonzero because $\delta_{\tilde{\mathbf{m}}}$ does not meet ∞. The formula implies that $\tilde{\mathbf{m}}$ and $\mathbf{m}$ are equivalent under a flow generated by the W_α and the Y^r:

$$\textbf{(3.10e)} \qquad \begin{cases} \tilde{\mathbf{x}} = \operatorname{diag}(f(r))^{-1} \, \mathbf{x} \operatorname{diag}(f(\infty_\alpha)) \\ \tilde{\mathbf{u}} = \operatorname{diag}(f(r)) \, \mathbf{u} \operatorname{diag}(f(\infty_\alpha))^{-1} . \end{cases}$$

(3.11) THEOREM: The isospectral surface $M(\mathcal{R})$ is a smooth $g_{\mathcal{R}} + \ell + \ell' - 1$ dimensional manifold. The tangent spaces $T_m M(\mathcal{R})$ are spanned by the $g_{\mathcal{R}}$ vectors $X^*_{k,\alpha}(\mathbf{m})$, the ℓ' vector fields $Y^r(\mathbf{m})$ and the ℓ vector fields $W_\alpha(\mathbf{m})$.

Proof: The Y^r and W_α generate simple scaling flows. They preserve the G^r_α and therefore they are tangential to $M(\mathcal{R})$. There is one linear dependence relationship among the them; namely,

$$\sum_{r=1}^{\ell'} Y^r = \sum_{\alpha=1}^{\ell} W_\alpha .$$

They span a subspace of dimension $\ell + \ell' - 1$. This with (3.9) proves that $M(\mathcal{R})$ has dimension at least $g_{\mathcal{R}} + \ell + \ell' - 1$. The divisor map is, by the discussion (3.10), well defined on the space of orbits of the scaling action generated Y^r and the W_α. It is an isomorphism of the orbit space onto the space of all positive divisors of degree $g_{\mathcal{R}}$ satisfying the condition (2.2a). See (3.6b). The Abel mapping is an isomorphism of the latter space of divisors onto an open affine subset of the Jacobian $\mathcal{J}(\mathcal{R})$. This proves that $M(\mathcal{R})$ is a smooth manifold of dimension $g_{\mathcal{R}} + \ell + \ell' - 1$. □

(3.12) THEOREM: The isospectral matrices $\mathbb{L}$ and Z satisfy these Lax equations:

$$X^*_{k,\alpha}(\mathbb{L}) = [\mathbb{B}, \mathbb{L}] \quad X^*_{k,\alpha}(Z) = [B_{k,\alpha}, Z]$$

globally on M.

Proof: The Lax equations were shown to hold on a particular integral curve in (2.22e) and (2.26e). A dense open set of solutions may be realized by the method of Chapter II. Therefore the Lax equations hold globally. □

CHAPTER IV

HAMILTONIAN FORMALISM

Preliminaries. The space $\mathbb{C}^{2\ell\ell'}$ is a symplectic mainifold with respect to the following 2 form:

$$(4.1a) \qquad \omega'_{\mathbf{m}} = \mathrm{tr}(d\mathbf{x} \wedge d\mathbf{u}^T) = \sum_{r=1}^{\ell'} \sum_{\alpha=1}^{\ell} dx^r_\alpha \wedge du^r_\alpha.$$

The Hamiltonian vector field X_f of an $f \in C^\infty(\mathbb{C}^{2\ell\ell'})$ is defined by this formula: $\omega'(X_f, \cdot) = df$. The Poisson bracket, $\{,\} : C^\infty(\mathbb{C}^{2\ell\ell'}) \times C^\infty(\mathbb{C}^{2\ell\ell'}) \to C^\infty(\mathbb{C}^{2\ell\ell'})$, associated to ω' is defined for $f, g \in C^\infty(\mathbb{C}^{2\ell\ell'})$ by this formula: $\{f,g\}(\mathbf{m}) = \omega'(X_f, X_g)$. These objects are given explicitely in the following formulas:

$$(4.1b) \qquad X_f(\mathbf{m}) = f_u\, \partial_x - f_x\, \partial_u$$

and

$$(4.1c) \qquad \{f,g\}(\mathbf{m}) = \mathrm{tr}(\partial_x f \cdot \partial_u g^T - \partial_x g \cdot \partial_u f^T)$$

where $\partial_x f$ is the $\ell' \times \ell$ matrix whose entry (r, α) is $\frac{\partial f}{\partial x^r_\alpha}$.

Let M be the algebraic submanifold of $\mathbb{C}^{2\ell\ell'}$ defined by the constraints (2.15a). The restiction of ω' to M, which shall be denoted ω, is defined for $\mathbf{m} \in M$ by this formula:

$$(4.1d) \qquad \omega_{\mathbf{m}} : T_{\mathbf{m}}(M) \times T_{\mathbf{m}}(M) \to \mathbb{C}, \quad \omega_{\mathbf{m}}(X,Y) = \omega'_{\mathbf{m}}(X,Y).$$

Nondegeneracy is not usually preserved under restriction. Thus it may not be possible to associate a Hamiltonian vector field with a function on $C^\infty(M)$.

Let $H \in C^{\infty}(\mathbb{C}^{2\ell\ell'})$. The Hamiltonian vector field X_H is not in general tangential to M. However, if ω is nondegenerate, there is a unique function $H^* \in C^{\infty}(\mathbb{C}^{2\ell\ell'})$ such that $H^*|_M = H|_M$ and the Hamiltonian vector field of H^* is tangential to M. H^* is called the *restricted Hamiltonian* [14]. The first condition suggests, from algebraic geometry, that $H^* - H$ belongs to the ideal generated by the constraints. This is indeed the case ([14] formula (5.3)).

Hamiltonians. It is clear that the matrix $\tilde{\mathbb{L}}$ is isospectral with respect to the AKNS/Neumann flows. This leads to the following definition.

(4.2) DEFINITION: The following formulas are used to define matrices Γ_j and $\Gamma_{k,j}$:

$$\tilde{\mathbb{L}} = \mathbf{a} + \mathbf{x}(\zeta^m z - \epsilon)^{-1}\mathbf{u}^T\mathbf{a}^m = \mathbf{a} + \sum_{j=1}^{\infty} \Gamma_\alpha \tilde{z}^{-j} \quad \text{where} \quad \Gamma_j = \mathbf{x}\epsilon^{j-1}\mathbf{u}^T\mathbf{a}^m$$

and

$$\tilde{\mathbb{L}}^{k-m\alpha} = \mathbf{a}^{k-m\alpha} + \sum_{j=1}^{\infty} \Gamma_{k,j}\tilde{z}^{-j}.$$

Let $H_{k,j}$ be the function of $\mathbf{m}$ defined by this formula:

$$H_k \doteq \frac{1}{k - m\alpha} \operatorname{Tr}(\mathbf{a}^{-m}\tilde{\mathbb{L}}^{k-m\alpha}) \doteq \sum_{j=0}^{\infty} H_{k,j}\tilde{z}^{-j}$$

where

$$H_{k,j} \doteq \frac{1}{k - m\alpha} \operatorname{Tr}(\mathbf{a}^{-m}\Gamma_{k,j}).$$

(4.3) THEOREM: The AKNS-Neumann vector field $X_{k,j}$ is Hamiltonian with respect to ω and the Hamiltonian is $H_{k+1,j+1}$. Hamilton's equations are given by these formulas:

$$\text{(4.3a)} \qquad \partial_{k,j}\,\mathbf{m} = X_{k,j}(\mathbf{m}) : \quad \begin{cases} \partial_{k,j}\,\mathbf{x} = \sum_{\beta=1}^{j+1} \Gamma_{k,j+1-\beta}\mathbf{x}\epsilon^{\beta-1} \\ \\ \partial_{k,j}\,\mathbf{u} = \sum_{\beta=1}^{j+1} \Gamma_{k,j+1-\beta}^T\mathbf{u}\epsilon^{\beta-1}. \end{cases}$$

Proof: Let Y denote the Hamiltonian vector field of $H_{k+1,\alpha+1}$. It is a vector field on $\mathbb{C}^{2\ell\ell'}$. Let $\partial_{\tilde{\mathbb{L}}} H$ be the $\ell' \times \ell'$ matrix whose entry (s,t) is $\partial H / \partial \tilde{\mathbb{L}}_{s,t}$. Let $\partial_{x^r_\alpha} \tilde{\mathbb{L}}$ be the $\ell' \times \ell'$ matrix whose entry (s,t) is $\partial_{x^r_\alpha} \tilde{\mathbb{L}}_{s,t}$. Then

$$\partial_{\tilde{\mathbb{L}}} H_{k+1} = \mathbf{a}^{-m}\tilde{\mathbb{L}}^{k-m\alpha} == \mathbf{a}^{-m}\left(\mathbf{a}^{k-m\alpha} + \sum_{j=1}^{\infty} \Gamma_{k,j}\tilde{z}^{-j}\right)$$

and the chain rule gives this formula:

(4.3b)
$$\begin{aligned}\partial_{u^r_\gamma} H_{k+1} &= Tr\left((\partial_{\tilde{\mathbb{L}}} H_{k+1})\cdot(\partial_{u^r_\gamma}\tilde{\mathbb{L}})\right)\\ &= \sum_{j=1}^{\infty} \mathrm{Tr}\left(\mathbf{a}^{-m}\sum_{\beta=1}^{j}\Gamma_{k,j-\beta}\,\partial_{u^r_\gamma}\Gamma_\beta\right)\tilde{z}^{-j}\end{aligned}$$

where $\Gamma_{k,0}\doteq\mathbf{a}^{k-m\alpha}$. Let $e_{\gamma,r}$ be the $\ell\times\ell'$ matrix with a 1 in entry (γ,r) and zeros elsewhere. The formula in the $\tilde{z}^{-j-1}$ term in (4.3b) and the formula in (4.2) for Γ_β imply this formula:

(4.3c)
$$\begin{aligned}\partial_{u^r_\gamma} H_{k+1,j+1} &= \mathrm{Tr}\left(\mathbf{a}^{-m}\sum_{\beta=1}^{j+1}\Gamma_{k,j+1-\gamma}\,\partial_{u^r_\gamma}\Gamma_\beta\right)\\ &= \mathrm{Tr}\left(\mathbf{a}^{-m}\sum_{\beta=1}^{j+1}\Gamma_{k,j+1-\beta}\mathbf{x}\epsilon^{\beta-1}e_{\gamma,r}\mathbf{a}^m\right).\end{aligned}$$

The right hand side of this formula aggrees with the right hand side of (4.3a) because the (r,γ) entry of a matrix A is given by $\mathrm{tr}(Ae_{\gamma,r})$. It also agrees with the right hand side of (2.28′) with (2.27′) under our definition (4.2) of the $\Gamma_{k,j}$. This proves that $Y = X_{k,j}$ and therefore $X_{k,j}$ is Hamiltonian. □

(4.3d) REMARK: The Hamiltonian vector field of $H_{k+1,j+1}$, as it agrees with $X_{k,j}$, is tangential to M.

(4.3e) REMARK: In view of (2.28), the matrix $\mathbb{L}$ in (2.26c) provide another set of Hamiltonians for the AKNS-Neumann hierarchy.

Symplectic Manifold. Let $\mathcal{M}^1_\ell = \mathcal{M}^1\cap\mathcal{M}_\ell$ where $\mathcal{M}^1$ and $\mathcal{M}_\ell$ were defined in Chapter I. Then $\mathcal{M}^1_\ell$ is an open dense subset of $\mathbb{C}^{2\ell\ell'}$ which is invariant under the $\mathbf{G}_\epsilon$ action defined under (1.6). Let us consider the following modification of that action:

(4.4a)
$$\exp(\Theta)\cdot\mathbf{m}\doteq(e^{a^b_r\Theta}x^r, e^{-a^b_r\Theta}u^r)_{r=1}^{\ell'}$$

where

$$b = m-1+\delta_{m,0} = \begin{cases} 0 & \text{if } m=0,\\ m-1 & \text{if } m>0,\end{cases}$$

and Θ is an $\ell \times \ell$ diagonal marix. This action is Hamiltonian and its momentum mapping is given by this formula:

$$J : \mathcal{M}_\ell^1 \to \mathfrak{g}_\epsilon, \quad J(\mathbf{m}) = \operatorname{diag}(\mathbf{a}^b x_\alpha \cdot u_\alpha)_{\alpha=1}^{\ell} \cdot = \operatorname{diag}(\mathbf{x}^T \mathbf{a}^b \mathbf{u}) \tag{4.4b}$$

(4.5) PROPOSITION: If $\mathbf{m} \in \mathcal{M}_\ell^1$ then $\nu \doteq J(\mathbf{m})$ is a regular value of J.

Proof: The mapping of tangent space: $T_m J : T_m \mathcal{M}_\ell^1 \to \mathfrak{g}_\epsilon$, which is given by this formula: $T_m J(v, w) = \operatorname{diag}(\mathbf{u}^T \mathbf{a}^b v + \mathbf{x}^T \mathbf{a}^b w)$, is surjective because $\mathbf{m} \in \mathcal{M}_\ell$. □

The value of ν is given in (2.15a) and (2.16b). If $m = 0$ then $\nu = z_1$ and if $m > 0$ then $\nu = \boldsymbol{\epsilon}$. Let $\mathcal{P}'_\nu = J^{-1}(\nu)/\mathbf{G}_\epsilon$. Then, since the $\mathbf{G}_\epsilon$ action is free and proper [3], $\mathcal{P}'_\nu$ is a manifold. Its dimension is given by this formula:

$$\dim(\mathcal{P}'_\nu) = 2\ell\ell' - 2\ell. \tag{4.6}$$

It is a Marsden-Weinstein reduced phase space. Therefore the natural mapping $\pi_\nu : J^{-1}(\nu) \to \mathcal{P}_\nu$ is a Poisson mapping [2]. The symplectic form ω'_ν satisfies this formula: $\pi_\nu^* \omega'_\nu = i_\nu^* \omega'$ where i_ν is the inclusion mapping of $J^{-1}(\nu)$ in $\mathcal{M}_\ell^1$.

The constraints (2.15a), being constant along $\mathbf{G}_\epsilon$ orbits, are functions on $\mathcal{P}'_\nu$. Let $\mathcal{P}_\nu$ be the set of orbits satisfying the constraints. If $m = 0$ then $\mathcal{P}_\nu = \mathcal{P}'_\nu$. The determinant of the $m\ell(\ell-1) \times m\ell(\ell-1)$ matrix whose entries are given for $\alpha \neq \alpha'$, $\beta \neq \beta'$ and $0 \leq j, j' \leq m-1$, by this formula:

$$\{\mathbf{a}^j x_\alpha \cdot u_\beta, \mathbf{a}^{j'} x_{\alpha'} \cdot u_{\beta'}\} = \mathbf{a}^{j+j'} x_{\alpha'} \cdot u_\beta \delta_{\alpha,\beta'} - \mathbf{a}^{j+j'} x_\alpha \cdot u_{\beta'} \delta_{\alpha',\beta},$$

is nondegenerate. According to ([14], §6), $\mathcal{P}_\nu$ is a symplectic manifold. Its dimension is given by this formula:

$$\dim(\mathcal{P}_\nu) = 2\ell\ell' - 2\ell - m\ell(\ell-1) = 2(g_{\mathcal{R}} + \ell' - 1), \tag{4.7}$$

where $g_{\mathcal{R}}$, the genus of the isospectral curve of $\mathbf{m}$ is given in (3.2).

The $\mathbf{H}_a$ action (1.8) preserves the constraints and it commutes with the $\mathbf{G}_\epsilon$ action. Therefore there is an induced $\mathbf{H}_a$ action on $\mathcal{P}_\nu$. This action is Hamiltonian and its momentum mapping is given as in the third formula in (1.8) by this formula:

$$J_1 : \mathcal{P}_\nu \to \mathfrak{h}_a, \quad J_1(\mathbf{G}_\epsilon \cdot \mathbf{m}) = \operatorname{diag}(x^r \cdot u^r)_{r=1}^{\ell'}.$$

(4.8) PROPOSITION: If $\mathbf{m} \in \mathcal{M}_\ell^1$ then $c \doteq J_1(\mathbf{G}_\epsilon \cdot \mathbf{m})$ is a regular value of J_1.

Proof: The mapping of tangent space: $T_m J_1 : T_m P_\nu \to \mathfrak{g}_\epsilon$, which is given by this formula: $T_m J_1(v,w) = \mathrm{diag}(u^r \cdot v^r + x^r \cdot w^r)$, is surjective because $\mathbf{m} \in M^1$. □

The value of c is is given in (2.20b). Let $\mathcal{P}_{\nu,c} = J_1^{-1}(c)/\mathbf{H}_a$. Then, since the $\mathbf{H}_a$ action is free and proper [3], $\mathcal{P}_{\nu,c}$ is a manifold. The components of J and J_1 are independent up to the following relation:

$$\sum_{\alpha=1}^{\ell} x_\alpha \cdot u_\alpha = \sum_{r=1}^{\ell'} x^r \cdot u^r .$$

Therefore the dimension of $\mathcal{P}_{\nu,c}$ is given by this formula:

$$\dim(\mathcal{P}_{\nu,c}) = 2g_{\mathcal{R}} . \tag{4.9}$$

It is a Marsden-Weinstein reduced phase space. Therefore the natural mapping, $\pi_c : J_1^{-1}(c) \to \mathcal{P}_{\nu,c}$, is a Poisson mapping [2]. The symplectic form $\omega_{\nu,c}$ satisfies this formula: $\pi_c^* \omega_{\nu,c} = i_c^* \omega_\nu$ where i_c is the inclusion mapping of $J_1^{-1}(c)$ in $\mathcal{P}_\nu$. The relationship between these symplectic manifolds is summarized in the following diagram:

$$\begin{array}{ccccc} & & & & J^{-1}(\nu) \xrightarrow{\ i_\nu\ } \mathcal{M}_\ell^1 \\ & & & & \downarrow{\scriptstyle \pi_\nu} \\ J_1^{-1}(c) & \xrightarrow{\ i_c\ } & (P_\nu, \omega_\nu) & \xrightarrow{\ i\ } & (P'_\nu, \omega'_\nu) \\ {\scriptstyle \pi_c}\downarrow & & & & \\ (P_{\nu,c}, \omega_{\nu,c}) & & & & \end{array}$$

More on Hamiltonians: Independence and Involution. It was shown using (1.10a) that the Hamiltonians $H_{k,j}$ are in involution on $\mathbb{C}^{2\ell\ell'}$. This was done using Lie algebraic considerations such as the Adler-Kostant-Symes theorem (1.2) and the Guillemin-Sternberg theorem (1.5). The Hamiltonians vector fields $X_{k,j}$ annihilate the entries of J. Therefore they are tangential to $J^{-1}(\nu)$. The Hamiltonians are constant along orbits of the $\mathbf{G}_\epsilon$ action (4.4a); hence, they drop down to functions on P'_ν. They are in involution because π_ν is a Poisson map.

The vector fields $\pi_{\nu *} X_{k,j}$ are tangential to P_ν. See (2.19), (3.9) and (4.3d). They therefore commute with respect to ω_ν.

The vector fields $X_{k,j}$ annihilate the entries of J_1. Therefore the $\pi_{\nu*}X_{k,j}$ are tangential to $J_1^{-1}(\nu)$. The Hamiltonians are constant along orbits of the $\mathbf{H}_a$ action (1.8); hence, they drop down to functions on $P_{\nu,c}$. They are in involution because π_ν is a Poisson map.

Let $\mathcal{P}_{\nu,c}(\mathcal{R})$ be the level set of the G^r_α in $\mathcal{P}_{\nu,c}$. By theorem (3.11) the $g_{\mathcal{R}}$ Hamiltonian vector fields $\pi_{c*} \circ \pi_{\nu*}X_{k,j}$ corresponding to the Weierstrass gap numbers of ∞ form a basis for the space of smooth vector fields on $\mathcal{P}_{\nu,c}(\mathcal{R})$. This proves that the AKNS/Neumann hierarchy is completely integrable.

Algebraic Complete Integrability. The formula in [18a] for the Baker function in terms of a Riemann theta function over $\mathcal{R}$ implies that the generalized Neumann systems are *algebraically* completely integrable Hamiltonian systems.

The divisor $\delta_{\mathbf{m}}, \mathbf{m} \in M(\mathcal{R})$, is well defined modulo the $\mathbf{H}_a$ action. See the discussion sorrounding (3.10a). The mapping $\mathcal{A}\circ\delta$, where $\mathcal{A}$ is the Abel mapping, is well defined mudulo the $\mathbf{G}_\epsilon$ action. See the discussion sorrounding (3.10b). It is an isomorphism of $\mathcal{P}_{\nu,c}(\mathcal{R})$ onto an affine subset $\mathcal{J}(\mathcal{R})\backslash\Theta$ See the discussion preceeding (2.3). The image of an integral curve of any member of our AKNS/Neumann hierarchy under $\mathcal{A} \circ \delta$ is a line in $\mathcal{J}(\mathcal{R})$.

Appendix: The Neumann System

The generalities of the introduction are best illustrated by the premier example, the Neuman system, which is given by these formulas:

$$\text{(A1)} \qquad \dot{\mathbf{m}} = X^*(\mathbf{m}) : \quad \begin{cases} \dot{x}_1 = x_2 \\ \dot{x}_2 = (\mathbf{a} + \sigma(\mathbf{x}, \mathbf{y}))x_1 \end{cases}$$

where $\mathbf{m} = (x_1, x_2)$ satisfies these *constraints*:

$$\text{(A2)} \qquad x_1 \cdot x_1 = 1, \qquad x_1 \cdot x_2 = 0,$$

and $\mathbf{q}$ is given by this *trace formula*:

$$\text{(A3)} \qquad \sigma = -\mathbf{a}x_1 \cdot x_1 - x_2 \cdot x_2,$$

$(x_1, x_2) \in \mathbb{R}^{2g+2}$, $\mathbf{a}$ is a constant diagonal matrix whose entries are distinct. The approach below is based on a method, developed originally by H. Flaschka [9], which generalizes the relationship between the Neumann system and Krichever's algebraic spectral theory of Hill's equation [12],

$$\text{(A4)} \qquad L\psi = \lambda\psi \quad \text{where} \quad L = \partial^2 - q(\tau).$$

The Neumann system may be regarded as a system of harmonic oscillators constrained to the tangent space TS^g to the g-dimensional sphere. See [8], [14] or [16] for details. The symplectic form $\omega = dx_1 \wedge dx_2$, restricted to TS^g is nondegenerate; hence, (TS^g, ω) is a symplectic manifold. The vector field X^* is Hamiltonian with respect to (TS^g, ω). See also [4a], [4b], [6], [7], [8], [14], [16] and [18b] for more on the Neumann system.

The are no constraints such as (A2) constraining the initial data in the AKNS-Neumann problem (0.6a). However, some of the systems defined later are constrained.

Finite Dimensional Invariant Surfaces. The Korteweg-de Vries equation, $q_t = \frac{1}{4}(6qq_{xx} + q_{xxx})$, is just one member of an infinite family of λ preserving deformations of L. Like the AKNS hierarchy, it has finite dimensional invariant surfaces. Krichever's approach [12] to these invariant sets uses commutative rings of differential operators. If $q(x,t)$ satisfies the Korteweg-de Vries and if $\partial^2 - q(x,0)$ commutes with an operator of odd order then $\partial^2 - q(x,t)$ commutes with an operator of odd order. An invariant set of the Korteweg-de Vries equation is identified in a concrete way with the Jacobian variety of a hyperelliptic curve. The setup parallels scattering theory. The scattering data, reflection coefficients, transmission coefficients and norming constants, is replaced by algebraic data, divisors on a hyperelliptic curve.

The Neumann system is another approach to these invariant sets. A solution $\mathbf{m}(\tau)$ to the Neumann problem (A1) corresponds to a differential operator P of order $2g+1$ that commutes with L. The genus of the spectral curve is g. Conversely, any operator L, as in (A4), that commutes with an operator of odd order corresponds in a unique way to a solution to the Neumann problem. Hamilton's equations are obviously explicit and, as we shall see, **(i)** there are explicit formulas for its constants of motion, **(ii)** a level set of the constants of motion is invariant under the Korteweg-de Vries flow and **(iii)** a level set of the constants of motion is isomorphic to an open affine subset of the Jacobian variety of a hyperelliptic curve. The invariant manifold, the Jacobian, is a level set of the constants of motion of the Neumann system.

The isospectral deformation theory of the Korteweg-de Vries equation plays a key role in the development of several aspects of the Neumann system; e.g., the complete integrabiltiy of the Neumann system. This has led to the following conjecture concerning the nature of completely integrable systems. *The level surfaces of the constants of motion are invariant surfaces of an integrable partial differential equation.* As the isospectral techniques for dealing with integrable partial differential equations are very well developed, such a theorem could provide a systematic way of finding constants of motion for a finite dimensional Hamiltonian system.

Isospectral Matrices. Let $x^r = (x_1^r, x_2^r)^T$ and $u^r = (-x_2^r, x_1^r)^T$ where $1 \leq r, s \leq g+1$ and let $x^r \otimes u^s$ be the 2×2 matrix whose entry (i,j) is $x_i^r u_j^s$. Let $\mathbf{m} = (\mathbf{x}, \mathbf{u})$. The method of solution to the initial value problem for the Neumann system is based on these formulas:

$$\textbf{(A5)} \qquad \dot{Z} = [B, Z] \quad \text{where} \quad \begin{cases} Z \doteq e_{2,1} + \displaystyle\sum_{r=1}^{g+1} \frac{x^r \otimes u^r}{\lambda - a_r} \\ B \doteq (\lambda + \sigma(\mathbf{m}))e_{2,1} + e_{1,2}. \end{cases}$$

The Lax equation in (A5) implies that the characteristic polynomial of Z is constant along any integral curve of the vector field X^* on TS^g. The characteristic polynomial is given in these formulas:

$$\textbf{(A6)} \qquad g(\lambda, z) \doteq |Z(\mathbf{m}, \lambda) - zI| = z^2 - \sum_{r=1}^{g+1} \frac{G^r}{\lambda - a_r}$$

where

$$G^r = (x_1^r)^2 - \sum_{s \neq r} \frac{(x_1^r x_2^s - x_1^s x_2^r)^2}{a_r - a_s}, \qquad (r = 1, \ldots g+1).$$

Let R be the algebraic curve: $g(\lambda, z) = 0$; R consists of 2 sheets merging at $2g+2$ branch points. There is a branch point at $\lambda = \infty$ denoted ∞. The rational function λ has a double pole at ∞ and z has a simple zero at ∞. The remaining $2g+1$ branch points are the zeros and poles of z away from the point ∞. The poles of z are points $(\lambda, z) = (a_r, \infty)$ denoted simply by a_r. Now z has g zeros other than ∞ denoted b_j by their λ coordinate. The divisor of z is given by

$$\textbf{(A7)} \qquad (z) = \infty + \sum_{j=1}^{g} b_j - \sum_{r=1}^{g+1} a_r.$$

The genus of R is g.

The G^r are of course constants of motion. Uhlenbeck (1975) and Devaney (1978) (see [14c], page 186) found these constants of motion evidently through an exhaustive search. The matrix Z, from which the G^r were just derived, was found in [18b] in a systematic way using the Riemann-Roch theorem for Baker functions. A Baker function is a function on a Riemann surface with prescribed poles and essential singularities. Krichever's Riemann-Roch theorem [18a] gives the dimension of certain linear spaces of Baker functions. The first use of Baker functions goes back to Krichever and Novikov (e.g. [12]).

The 2-form $\omega(\mathbf{m})$, $\mathbf{m} \in TS^g$, determines the following natural projection:

$$T_{\mathbf{m}}\mathbb{R}^{2g+2} \to T_{\mathbf{m}}TS^g.$$

Let G^{r*} denote the Hamiltonian for the projection to TS^g of the Hamiltonian vector field of G^r. These Hamiltonians form a set of g independent involutive constants of motion. This may be proved by direct but incredibly tedious computation. A geometrical proof of their involutivity using a loop algebra is given below. The connection between the Neumann system and the Korteweg-de Vries equation is used below to give a proof of their independence.

The Solution. The spectral problem $Z(\lambda, \mathbf{m})U(p) = z(p)U(p)$ has a unique solution of the form $U(p) = (1, f(p))$ where f is a rational function with a simple pole at ∞ and g finite poles; say, $\delta = \delta_1 + \cdots + \delta_g$. The divisor δ is nonspecial. By the Riemann-Roch theorem [18a], there exists a unique function $\psi(\tau, p)$, *the Baker function*, satisfying these conditions:

(*i*) Any pole of ψ lies in δ and

(*ii*) $\psi(\tau, p) = (1 + O(\kappa^{-1}))\exp(\kappa\tau)$ at ∞ where $\kappa \doteq \sqrt{\lambda}$.

The Baker function ψ satisfies an operator spectral problem of the form (A4); equivalently,

$$\textbf{(A8)} \qquad \dot{U}(\tau, p) = B(\lambda, \mathbf{m}(\tau))U(\tau, p) \quad \text{where} \quad B = (\lambda + q)e_{2,1} + e_{1,2}$$

and $U(\tau, p) \doteq (\psi, \dot{\psi})^T$. The solution $\mathbf{m}(\tau)$ is obtained by evaluating the Baker function along the polar divisor of z:

$$\textbf{(A9)} \qquad x_1^r(\tau) = c_r\psi(a_r, \tau) \quad \text{and} \quad x_2^r(\tau) = c_r\dot{\psi}(a_r, \tau)$$

where $c_r = x_1^r(0)$. The proof that this gives the solution to (IVP) follows from a series of residue calculations involving abelian differentials on R. These calculations are analogous to the Deift-Lund-Trubowitz *trace formulas* [8]. Here, *trace formulas* show that the point $\mathbf{m}(\tau)$ defined by (A10) lies on TS^g and that

$$\textbf{(A10)} \qquad q(\tau) = \sigma(\mathbf{m}(\tau))$$

where σ is given in (0.14). These facts and (A8) with $\lambda = a_r$ prove that $\mathbf{m}$ satisfies (IVP).

The Baker function ψ is expressible in terms of the Riemann θ function of the Riemann surface R [18a]. This establishes the algebraic nature of the solution.

One of the most interesting results of this project is a proof [18b] that U satisfies the following spectral problem:

$$\textbf{(A11)} \qquad Z(\mathbf{m}(\tau), \lambda(p))U(\tau, p) = z(p)U(\tau, p)$$

The Lax equation (A5) is the integrability condition underlying this equation and (A8).

J. Moser [14] discovered the following relationship between the Neumann system and the geometry of quadrics. It seems that he was motivated to search for a relationship based on the following 2 facts: **(i)** The solution to the geodesic problem on an ellipsoid is given by hyperelliptic functions and **(ii)** The geodesic problem has a Lax equation that is similar to a Lax equation of the Neumann problem. The following approach uses the isospectral matrix Z. It is based solely on isospectrality and so it reaffirms the connection established by Moser between the geometry of quadrics and the Lax idea.

(A12) THEOREM: Let $\mathbf{m}(\tau) = (x_1(\tau), x_2(\tau))$ be a solution to the Neumann problem and let $l(\tau)$ be the line in $\mathbb{R}^{g+1}$ through $x_2(\tau)$ in the direction $x_1(\tau)$:

$$l(\tau) : x_2(\tau) + s x_1(\tau) \quad \text{where} \quad -\infty < s < \infty.$$

There exists g **τ-independent** confocal quadrics $Q_j \quad (j = 1, \cdots, g)$ in $\mathbb{R}^{g+1}$ such that $l(\tau)$ is tangential to each Q_j.

Proof: Let us consider the family of confocal quadrics defined by this formula:

$$Q_\lambda \doteq \sum_{r=1}^{g+1} \frac{Y_r^2}{\lambda - a_r} = 1,$$

and the g finite branch points b_j of R where $z = 0$. Let $Q_j = Q_{b_j}$. Equation (8) with $p = b_j$ is equivalent to these equations:

$$\text{(A13)} \qquad \sum_{r=1}^{g+1} \frac{x_1^r\left(x_2^r - \frac{\dot\psi(\tau,b_j)}{\psi(\tau,b_j)} x_1^r\right)}{b_j - a_r} = 0$$

and

$$\text{(A14)} \qquad \sum_{r=1}^{g+1} \frac{x_2^r\left(x_2^r - \frac{\dot\psi(\tau,b_j)}{\psi(\tau,b_j)} x_1^r\right)}{b_j - a_r} = 1.$$

These equations combined imply this one:

$$\text{(A15)} \qquad \sum_{r=1}^{g+1} \frac{\left(x_2^r - \frac{\dot\psi(\tau,b_j)}{\psi(\tau,b_j)} x_1^r\right)^2}{b_j - a_r} = 1.$$

Now (A15) is equivalent to this formula:

$$\textbf{(A16)} \qquad Q_j(\xi_j(\tau)) = 1 \quad \text{where} \quad \xi_j(\tau) \doteq x_2(\tau) - t(\tau) x_1(\tau)$$

and $t(\tau) = \partial \ln \psi(\tau, b_j)$. The equation (A13) implies that the line $l(\tau)$ is tangential to Q_j at $\xi_j(\tau)$; that is,

$$\textbf{(A17)} \qquad \nabla Q_j(\xi_j(\tau)) \cdot x_1 = 0. \quad \square$$

(A18) REMARK: In addition, Moser showed that $\xi_j(\tau)$, the point of contact, moves in τ along a geodesic on Q_j.

Independence. The formula (A8) establishes a relationship between the Neumann system and Krichever's operator theory. This relationship may be developed a bit further to prove that the Hamiltonians G^{r*} are independent. There exists for each $j \in \mathbb{N}$ a unique monic differential operator L_j of order j in ∂ such that

$$\textbf{(A19)} \qquad L_j \psi \cdot \psi^{-1} = \kappa^j + O(\kappa^{-1}).$$

It follows then that the operator $[L_j, L]$ has order 0. The Korteweg-de Vries hierarchy is the family of partial differential equations given by these operator Lax equation:

$$\textbf{(A20)} \qquad \frac{\partial L}{\partial t_j} = -\partial_j q = [L_j, L] \text{ for each } j \in \mathbb{N}.$$

The Neumann hierarchy is the family of nonlinear ordinary differential equations in $\mathbf{m}$ given by these formulas:

$$\textbf{(A21)} \qquad \frac{\partial x_1^r}{\partial t_j} = L_j x_1^r \quad \text{and} \quad \frac{\partial x_2^r}{\partial t_j} = \partial \left(\frac{\partial x_1^r}{\partial t_j} \right).$$

The system defines a vector field X_j^* which is tangential to TS^g. These vector fields are Hamiltonian with respect to the restriction of ω to TS^g. The Hamiltonian is a linear combination of the G^{r*}.

The Baker function ψ may be viewed as a function of $\mathbf{m}$. The action of the vector field X_j^* on ψ is given by $X_j^* \psi = L_j \psi$. A linear dependence relationship among the X_{2j-1}^*, $(j = 1, \ldots, g)$ would correspond to a nontrivial linear dependence relationship among the $L_{2j-1}\psi$, $(j = 1, \ldots, g)$. This is impossible by (A19). It follows then that vector fields X_{2j-1}^*, $(j = 1, \ldots, g)$ are linearly independent. This in turn implies that the G^{r*}, $(r = 1, \ldots, g)$ are functionally independent.

Involutivity. Let $\mathfrak{g} = gl(2,\mathbb{C})$, the Lie algebra of all 2×2 matrices over $\mathbb{C}$. The dual of $\mathfrak{g}$ is identified with $\mathfrak{g}$ itself using the usual trace form $(\xi,\eta)\in\mathfrak{g}\times\mathfrak{g}\to \mathrm{tr}(\xi\eta)$. The *loop algebra* $L(\mathfrak{g})$ in the parameter λ is the infinite dimensional Lie algebra defined by the following formulas:

$$L(\mathfrak{g}) = \{\,\xi(\lambda) = \lambda^N(\sum_{j=0}^{\infty}\xi^{[j]}\lambda^{-j}) : \xi^{[j]}\in\mathfrak{g} \quad\text{and}\quad N\in\mathbb{Z}\} \tag{A22}$$

$$[\xi(\lambda),\eta(\lambda)] = \lambda^{N+N'}\sum_{j=0}^{\infty}\left(\sum_{l+k=j}[\xi^{[l]},\eta^{[k]}]\right)\lambda^{-j}. \tag{A23}$$

Let K denote the bilinear form on $L(\mathfrak{g})$ given for $\xi(\lambda)$ and $\eta(\lambda)$ in $L(\mathfrak{g})$ by this formula:

$$K(\xi(\lambda),\eta(\lambda)) = \pi_{-1}\circ\mathrm{tr}(\xi(\lambda)\eta(\lambda)) = -\ \mathrm{Res}_{\infty}\,\mathrm{tr}(\xi(\lambda)\eta(\lambda))d\lambda. \tag{A24}$$

Then K is symmetric, nondegenerate and associative. Let us consider the direct sum decomposition of $L(\mathfrak{g})$ into these subalgebras:

$$L(\mathfrak{g}) = \mathfrak{N}\oplus\mathfrak{K} \tag{A25}$$

where $\mathfrak{N}$ is the subalgebra of polynomial loops, and

$$\mathfrak{K} = \mathbb{C}(e_{2,1}+e_{1,2}\lambda^{-1},\ e_{2,1}\lambda^{-1},\ e_{1,1}\lambda^{-1},\ e_{2,2}\lambda^{-1})\oplus\sum_{j=2}^{\infty}\mathfrak{g}\lambda^{-j}. \tag{A26}$$

The projections to $\mathfrak{N}$ and $\mathfrak{K}$ are given by these formula:

$$\begin{cases} \pi_{\mathfrak{N}}(\xi(\lambda)) = \pi_{\geq}(\xi(\lambda)) - \xi_{1,2}^{[-1]}e_{2,1}. \\ \pi_{\mathfrak{K}}(\xi(\lambda)) = \pi_{<}(\xi(\lambda)) + \xi_{1,2}^{[-1]}e_{2,1}. \end{cases} \tag{A27}$$

Let $\epsilon = 2e_{2,1}\in\mathfrak{N}^0\cap[\mathfrak{K},\mathfrak{K}]^0$, an infinitesimal character of $\mathfrak{N}$. The dual of $\mathfrak{N}$ is identified with the annihilator $\mathfrak{K}^0$ of $\mathfrak{K}$ with respect to K; $\mathfrak{N}^*\cong\mathfrak{K}^0$. The annihilator is given by this formula:

$$\mathfrak{K}^0 = \mathbb{C}(e_{2,1}-e_{1,2}\lambda^{-1},\ e_{2,1}\lambda^{-1},\ e_{1,1}\lambda^{-1},\ e_{2,2}\lambda^{-1})\oplus\sum_{j=2}^{\infty}\mathfrak{g}\lambda^{-j}. \tag{A28}$$

It is a well known fact that the dual of a Lie algebra is a Poisson manifold with respect to the Kirrillov-Poisson bracket. The space $\epsilon + \mathfrak{K}^0$ is a Poisson manifold under the identification (A28). By the Adler-Kostant-Symes Theorem, the following Hamiltonians are in involution with respect to the Kirrillov-Poisson bracket on $\epsilon + \mathfrak{K}^0$:

$$h_{i,j}(\epsilon + \xi(\lambda)) \doteq \operatorname{tr}\left(\lambda^i(\epsilon + \xi(\lambda))^j_{-1}\right) \quad \text{if} \quad \xi(\lambda) \in \mathfrak{K}^0.$$

The family of differential equations on $\epsilon + \mathfrak{K}^0$ defined by the Hamiltonians $h_{i,j}$ in (0.38) is equivalent to the Korteweg-de Vries hierarchy.

The obscure point in this construction is the decomposition (A25) of $L(\mathfrak{g})$. It might be motivated in the following way. The isospectral matrix Z belongs to $\epsilon + \mathfrak{K}^0$. There is much more. The Hamiltonian vector field D_h on $\epsilon + \mathfrak{K}^0$ associated to the invariant Hamiltonian

$$h(\xi(\lambda)) \doteq \frac{1}{2} K(\lambda\xi(\lambda), \xi(\lambda))$$

is given by the following formula:

$$D_h(\epsilon + \xi(\lambda)) = [\pi_{\mathfrak{N}}(\lambda(\epsilon + \xi(\lambda)), \epsilon + \xi(\lambda)].$$

If $\xi(\lambda) = \xi^{[1]}_{1,2} e_{2,1} + \xi^{[1]}\lambda^{-1} + \xi^{[2]}\lambda^{-2} + \cdots$ then

$$\pi_{\mathfrak{N}}\lambda(\epsilon + \xi(\lambda)) = (\lambda + \xi^{[1]}_{1,1} - \xi^{[2]}_{1,2})e_{2,1} + e_{1,2}.$$

The matrix B in (A8) is given by this formula:

$$B = \pi_{\mathfrak{N}}(\lambda Z)$$

and the basic linear equation (A8) has the form

$$\partial U(\tau, p) = \pi_{\mathfrak{N}}(\lambda Z) U(\tau, p).$$

Let $\mathcal{M}$ denote the cotangent bundle to the space of all $n \times \ell$ matrices. Let $\mathbf{x}$ denote an $2 \times \ell$ matrix with columns x^r and rows x_α:

$$\mathbf{x} = (x^1, \ldots, x^\ell) = (x_1^T, \ldots, x_n^T)^T.$$

An element of $\mathcal{M}$ has the form $\mathbf{m} = (\mathbf{x}, \mathbf{u})$ where $\mathbf{u}$ has rows u^r and columns u_α. The Lie algebra $\mathfrak{N}$ acts on $\mathcal{M}$. The action and its momentum mapping $\Upsilon : \mathcal{M} \to \mathfrak{K}^0$ are given for $\xi(\lambda) \in \mathfrak{N}$ by this formula:

(A29) $$\xi(\lambda) \cdot \mathbf{m} = (\xi(a_r)x^r, -\xi(a_r)^T u^r)$$

(A30) $$\Upsilon(\mathbf{m}) = \pi_{\mathfrak{K}^0}(\mathbf{x}(\lambda - \mathbf{a})^{-1}\mathbf{u}^T) = -e_{2,1} + \sum_{r=1}^{\ell} \frac{x^r \otimes u^r}{\lambda - a_r}.$$

The momentum map Υ is equivariant with respect to the infinitesimal action of $\mathfrak{N}$ on $\mathfrak{N}^*$:

$$\begin{array}{ccc} \mathcal{M} & \xrightarrow{\Upsilon} & \mathfrak{N}^* \cong \mathfrak{K}^0 = \mathfrak{K} \\ {\scriptstyle \xi(\lambda)}\big\downarrow & & \big\downarrow{\scriptstyle -\pi_{\mathfrak{K}^0} \circ \mathrm{ad}_{\xi(\lambda)}} \\ \mathcal{M} & \xrightarrow{\Upsilon} & \mathfrak{N}^* \cong \mathfrak{K}^0 = \mathfrak{K}. \end{array}$$

The isospectral matrix Z satisfies

(A31) $$Z = \epsilon + \Upsilon(\mathbf{m}) \in \epsilon + \mathfrak{K}^0.$$

By a theorem of Guillemin and Sternberg, any momentum mapping is a canonical mapping. It follows then that the Hamiltonians defined by this formula:

(A32) $$H_{i,j} \doteq h_{i,j}(\Upsilon(\mathbf{m})) = \mathrm{tr}(\lambda^i Z^j)_{-1}$$

are involutive. The G^r, being functions of the $H_{i,j}$, are involutive. Since the G^{r*} agree with the G^r on TS^g, the G^{r*} are involutive.

This completes the proof of the algebraic complete integrability of the Neumann system. A connection with the Korteweg-de Vries hierarchy was established and then used to prove that the Uhlenbeck constants of motion are independent. A connection with loop algebras was established and then used to prove that the Uhlenbeck constants of motion are in involution.

This approach to the involutivity of the Uhlenbeck constants of motion is based on the geometry of momentum mappings developed in [3]; in particular, the infinitesimal $\mathfrak{N}$ action that lead to the interpretation (A31) of Z in terms of a momentum mapping. The theory in [3] does not include the Neumann system. The loop algebra decomposition used there is more directly related to systems of AKNS type [10].

Bibliography

[1] Ablowitz, M. J., D. J. Kaup, A. C. Newell and H. Segur, The Inverse Scattering Transform - Fourier Analysis for Nonlinear Problems, Studies in Applied Mathematics, LIII, 4 (1974).

[2] Abraham, J. and J. Marden, Foundations of Mechanics, 2nd ed., Benjamin, New York (1978).

[3] Adams, M. R., J. Harnad and E. Previato, Isospectral Hamiltonian Systems in Finite and Infinite Dimensions I, Generalized Moser Systems and Moment Maps into Loop Algebras, Commun. Math. Phys. 117(1988)451-500.

[4] Adler, M. and P. van Moerbeke,

(a) Completely Integrable Systems, Euclidean Lie Algebras and Curves, Advances in Mathematics 38 (1980) 267–317.

(b) Linearization of Hamiltonian Systems, Jacobi Varieties and Representation Theory, Advances in Mathematics 38 (1980) 318–379.

(c) The Algebraic Integrability of Geodesic Flow on S0(4), Invent. Math. 67 (1982) 297–331.

[5] Campbell, D. K. and A. R. Bishop, Soliton Excitations in Polyacetylene and relativistic Field Theory Models, Nuclear Physics B200 [FS4](1982).

[6] Cherednik, I. V., Differential Equations for the Baker-Akhiezer Functions of Algebraic Curves, Funkts. Anal. Prilozh. 12 (3) (1978) 45–54.

[7] Deift, P. and E. Truibowitz, Inverse Scattering on the Line, CPAM, Vol. XXXII, 121–251 (1979).

[8] Deift, P., F. Lund and E. Trubowitz, Nonlinear Wave Equations and Constrained Oscillator Systems, Comm. Math. Phys. 74 (1980)141–188.

[9] Flaschka, H.,

(a) Relations Between Infinite and Finite Dimensional Isospectral Equations, Proc. RIMS Symposium on Nonlinear Integrable Systems-Classical and Quantum Theory, World Science Publishing Co. (1981).

(b) Toward an Algebro-Geometrical Interpretation of the Neumann System, Tohoku Math. J., 36, 3 (1984) 407–426.

(c) Construction of Conservation Laws for Lax Equations: Comments on a Paper by G. Wilson, Quart. J. Math. Oxford (2), 34 (1983) 61-65.

[10] Flaschka, H., A. C. Newell and T. Ratiu, Kac-Moody Algebras and Soliton Equations II and III, Physica 9D(1983), 300–323 and 324–332.

[11] Guillemin, V. and S. Sternberg, The Moment Map and Collective Motion, Annals of Physics 127 (1980) 220–253.

[12] Krichever, I. M., Integration of Nonlinear Equations by Methods of Algebraic Geometry, Functional Anal. Appl. 11 (1977) 12–26.

[13] McKean, H. p. Jr.,

(a) The Sine-Gorden and Sinh-Gorden Equations on the Circle, Comm. on Pure and Appl. Math. Vol XXXIV (1981) 197-257.

(b) Boussinesq's Equation on the Circle, Comm. Pure and Appl. Math. Vol. XXXIV (1981) 599-691.

[14] Moser, J.,

(a) Various Aspects of Integrable Hamiltonian Systems, in Proc. CIME Conference, Bressanone, Italy, June 1978, Prog. Math. 8, Birkhauser.

(b) Geometry of Quadrics and Spectral Theory, Chern Symposium (1979), Springer-Verlag (1980) 147–188.

(c) Integrable Hamiltonian Systems and Spectral Theory, Lezioni Fermiane, Pisa (1981).

[15] Previato, E., Hyperelliptic Quasi-Periodic and Soliton Solutions of the Nonlinear Schrödinger Equation, Duke Math. J., Vol. 52, No. 2 (1985) 329–377.

[16] Ratiu, T.,

(a) The C. Neumann Problem as a Completely Integrable System on an Adjoint Orbit, Trans. Amer. Math. Soc. 264 (1981) 321–329.

(b) Complete Integrability of the Rosochatius System, A. Inst. Phy., Conf. Proc. 88 (1982) 109–115.

(c) Involution Theorems, Lecture Notes in Math. 755, Springer-Verlag (1980) 219–257.

[17] Reyman, A. G. and M. A. Semenov-Tian-Shansky,

(a) Reduction of Hamiltonian Systems, Affine Lie Algebras and Lax Equations, Inventiones Math. 58, (1979) 81-100.

(b) Reduction of Hamiltonian Systems, Affine Lie Algebras and Lax Equations II, Inventiones Math. 63 (1981) 423–432.

[18] Schilling, R. J.,

(a) Baker Functions for Compact Riemann Surfaces, Proc. of the AMS, Vol. 98 No. 4, (1986) 671-675.

(b) Generalizations of the Neumann System, Comm. on Pure and Appl. Math. Vol. XL (1987) 425-522.

University of Arkansas at Little Rock
Little Rock, Arkansas 72204
E-MAIL: RJSCHILLING@UALR.EDU

Editorial Information

To be published in the *Memoirs*, a paper must be correct, new, nontrivial, and significant. Further, it must be well written and of interest to a substantial number of mathematicians. Piecemeal results, such as an inconclusive step toward an unproved major theorem or a minor variation on a known result, are in general not acceptable for publication. *Transactions* Editors shall solicit and encourage publication of worthy papers. Papers appearing in *Memoirs* are generally longer than those appearing in *Transactions* with which it shares an editorial committee.

As of March 1, 1992, the backlog for this journal was approximately 9 volumes. This estimate is the result of dividing the number of manuscripts for this journal in the Providence office that have not yet gone to the printer on the above date by the average number of monographs per volume over the previous twelve months. (There are 6 volumes per year, each containing about 3 or 4 numbers.)

A Copyright Transfer Agreement is required before a paper will be published in this journal. By submitting a paper to this journal, authors certify that the manuscript has not been submitted to nor is it under consideration for publication by another journal, conference proceedings, or similar publication.

Information for Authors

Memoirs are printed by photo-offset from camera copy fully prepared by the author. This means that the finished book will look exactly like the copy submitted.

The paper must contain a *descriptive title* and an *abstract* that summarizes the article in language suitable for workers in the general field (algebra, analysis, etc.). The *descriptive title* should be short, but informative; useless or vague phrases such as "some remarks about" or "concerning" should be avoided. The *abstract* should be at least one complete sentence, and at most 300 words. Included with the footnotes to the paper, there should be the 1991 *Mathematics Subject Classification* representing the primary and secondary subjects of the article. This may be followed by a list of *key words and phrases* describing the subject matter of the article and taken from it. A list of the numbers may be found in the annual index of *Mathematical Reviews*, published with the December issue starting in 1990, as well as from the electronic service e-MATH [**telnet e-MATH.ams.com** (or **telnet 130.44.1.100**). Login and password are **e-math**]. For journal abbreviations used in bibliographies, see the list of serials in the latest *Mathematical Reviews* annual index. Authors are encouraged to supply electronic addresses when available. These will be printed after the postal address at the end of each article.

Electronically-prepared manuscripts. The AMS encourages submission of electronically-prepared manuscripts in $\mathcal{A}\mathcal{M}\mathcal{S}$-TEX or $\mathcal{A}\mathcal{M}\mathcal{S}$-LATEX. To this end, the Society has prepared "preprint" style files, specifically the amsppt style of $\mathcal{A}\mathcal{M}\mathcal{S}$-TEX and the amsart style of $\mathcal{A}\mathcal{M}\mathcal{S}$-LATEX, which will simplify the work of authors and of the production staff. Those authors who make use of these style files from the beginning of the writing process will further reduce their own effort.

Guidelines for Preparing Electronic Manuscripts provide additional assistance and are available for use with either $\mathcal{A}\mathcal{M}\mathcal{S}$-TEX or $\mathcal{A}\mathcal{M}\mathcal{S}$-LATEX. Authors with FTP access may obtain these *Guidelines* from the Society's Internet node e-MATH.ams.com (130.44.1.100). For those without FTP access they can be obtained free of charge from the e-mail address guide-elec@math.ams.com (Internet) or from the Publications Department, P. O. Box 6248, Providence, RI 02940-6248. When requesting *Guidelines* please specify which version you want.

Electronic manuscripts should be sent to the Providence office only after the paper has been accepted for publication. Please send electronically prepared manuscript files via e-mail to pub-submit@math.ams.com (Internet) or on diskettes to the Publications Department address listed above. When submitting electronic manuscripts please be sure to include a message indicating in which publication the paper has been accepted.

For papers not prepared electronically, model paper may be obtained free of charge from the Editorial Department at the address below.

Two copies of the paper should be sent directly to the appropriate Editor and the author should keep one copy. At that time authors should indicate if the paper has been prepared using $\mathcal{A}\mathcal{M}\mathcal{S}$-TEX or $\mathcal{A}\mathcal{M}\mathcal{S}$-LATEX. The *Guide for Authors of Memoirs* gives detailed information on preparing papers for *Memoirs* and may be obtained free of charge from AMS, Editorial Department, P.O. Box 6248, Providence, RI 02940-6248. The *Manual for Authors of Mathematical Papers* should be consulted for symbols and style conventions. The *Manual* may be obtained free of charge from the e-mail address cust-serv@math.ams.com or from the Customer Services Department, at the address above.

Any inquiries concerning a paper that has been accepted for publication should be sent directly to the Editorial Department, American Mathematical Society, P. O. Box 6248, Providence, RI 02940-6248.

Recent Titles in This Series

(Continued from the front of this publication)

435 **Aloys Krieg,** Hecke algebras, 1990
434 **François Treves,** Homotopy formulas in the tangential Cauchy-Riemann complex, 1990
433 **Boris Youssin,** Newton polyhedra without coordinates Newton polyhedra of ideals, 1990
432 **M. W. Liebeck, C. E. Praeger, and J. Saxl,** The maximal factorizations of the finite simple groups and their automorphism groups, 1990
431 **Thomas G. Goodwillie,** A multiple disjunction lemma for smooth concordance embeddings, 1990
430 **G. M. Benkart, D. J. Britten, and F. W. Lemire,** Stability in modules for classical Lie algebras: A constructive approach, 1990
429 **Etsuko Bannai,** Positive definite unimodular lattices with trivial automorphism groups, 1990
428 **Loren N. Argabright and Jesús Gil de Lamadrid,** Almost periodic measures, 1990
427 **Tim D. Cochran,** Derivatives of links: Milnor's concordance invariants and Massey's products, 1990
426 **Jan Mycielski, Pavel Pudlák, and Alan S. Stern,** A lattice of chapters of mathematics: Interpretations between theorems, 1990
425 **Wiesław Pawłucki,** Points de Nash des ensembles sous-analytiques, 1990
424 **Yasuo Watatani,** Index for C^*-subalgebras, 1990
423 **Shek-Tung Wong,** The meromorphic continuation and functional equations of cuspidal Eisenstein series for maximal cuspidal subgroups, 1990
422 **A. Hinkkanen,** The structure of certain quasisymmetric groups, 1990
421 **H. W. Broer, G. B. Huitema, F. Takens, and B. L. J. Braaksma,** Unfoldings and bifurcations of quasi-periodic tori, 1990
420 **Tom Lindstrøm,** Brownian motion on nested fractals, 1990
419 **Ronald S. Irving,** A filtered category $\mathcal{O}_S$ and applications, 1989
418 **Hiroaki Hijikata, Arnold K. Pizer, and Thomas R. Shemanske,** The basis problem for modular forms on $\Gamma_0(N)$, 1989
417 **John B. Conway, Domingo A. Herrero, and Bernard B. Morrel,** Completing the Riesz-Dunford functional calculus, 1989
416 **Jacek Nikiel,** Topologies on pseudo-trees and applications, 1989
415 **Amnon Neeman,** Ueda theory: Theorems and problems, 1989
414 **Nikolaus Vonessen,** Actions of linearly reductive groups on affine PI-algebras, 1989
413 **Francesco Brenti,** Unimodal log-concave and polya frequency sequences in combinatorics, 1989
412 **T. Levasseur and J. T. Stafford,** Rings of differential operators on classical rings of invariants, 1989
411 **Matthew G. Brin and T. L. Thickstun,** 3-manifolds which are end 1-movable, 1989
410 **P. L. Robinson and J. H. Rawnsley,** The metaplectic representation, Mp^c structures and geometric quantization, 1989
409 **William Arveson,** Continuous analogues of Frock space, 1989
408 **Idun Reiten and Michel Van den Bergh,** Two-dimensional tame and maximal orders of finite representation type, 1989
407 **Fokko du Cloux,** Représentations de longueur finie des groupes de Lie résolubles, 1989
406 **H. Toruńczyk and J. West,** Fibrations and bundles with Hilbert cube manifold fibers, 1989
405 **J. N. Brooks and J. B. Strantzen,** Blaschke's rolling theorem in $\mathbb{R}^n$, 1989

(See the AMS catalogue for earlier titles)

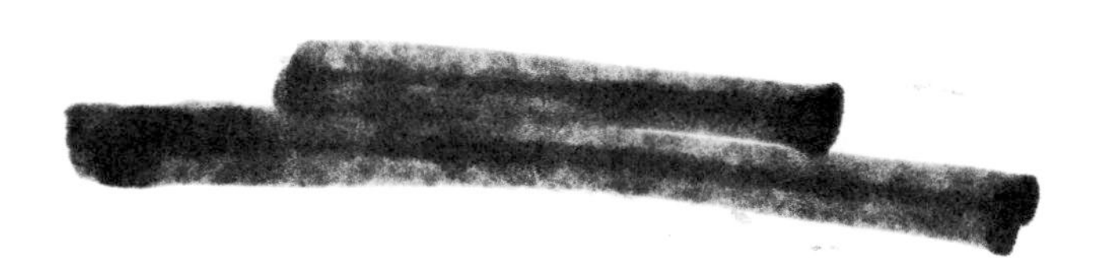